An Introduction to Structured Population Dynamics

J. M. Cushing
University of Arizona
Tucson, Arizona

An Introduction to Structured Population Dynamics

siam.
SOCIETY FOR INDUSTRIAL AND APPLIED MATHEMATICS

PHILADELPHIA

10 9 8 7 6 5 4 3 2 1

Library of Congress Cataloging-in-Publication Data

Cushing, J. M. (Jim M.), 1942-
An introduction to structured population dynamics / J.M. Cushing.
p. cm. -- (CBMS-NSF regional conference series in applied mathematics ; 71)
Outgrowth of a series of lectures given at a conference held at North Carolina University, Raleigh, during June of 1997.
Includes bibliographical references (p.) and index.
ISBN 0-89871-417-6 (pbk.)
1. Population biology--Mathematical models. I. Title.
II. Series.
QH352.C87 1998
577.8'8'015118--dc21 98-19033

To Alina and Lara

Contents

Preface

Interest in the dynamics of biological populations is quite old, its roots being traceable to the dawn of civilization. Many illustrious names are associated with early mathematical theories of population growth, e.g., Fibonnaci, Euler, Halley, Malthus [382]. The flowering of mathematical ecology and population dynamics occurred, however, during the first half of the twentieth century [245], [374]. Many of the "classical" equations (and their many variants) resulting from this development, such as the famous logistic equations, Volterra predator–prey equations, and the Lotka–Volterra competition equations, have had a tremendous influence on both theoretical and applied ecology and population dynamics. They stimulated the formulation of, and gave theoretical support to, many (if not most) of the fundamental tenets held today. These include exponential growth, carrying capacity, competitive exclusion, ecological niche, limiting similarity, r-K selectors, and predator–prey oscillations.

In order to gain a better understanding of the dynamics of biological populations, theoretical biologists and applied mathematicians have, over the course of the century, modified classical models and modeling methodologies in many ways. All mathematical models make simplifying assumptions, of course, and there is a relentless trade-off between biological accuracy and mathematical tractability. One way to view many of the simplifying assumptions made in population models is with regard to various uniformities and homogeneities that are either explicitly or implicitly postulated. For example, two common simplifications concern homogeneities in space and time. In such models there is no attempt to account for the spatial extent or movement of individuals or populations, and environments are assumed constant in time. This is the case in the classical models mentioned above. There exists now, however, a rather large and growing body of literature on spatial diffusion models for biological populations. There is also a growing body of literature on models that include either deterministic or stochastic environmental fluctuations.

Another important modeling assumption that is commonly made concerns the homogeneity of individuals within a population. Mathematical models often involve equations for total population "size" (total number or density of individuals or their total biomass, dry weight, etc.) and in effect treat all individuals within the population as identical. This is true of the classical models mentioned above and, indeed, of the majority of models studied today. However, individuals in biological populations differ with regard to their physiological and behavioral characteristics and therefore in the way they interact with their environment. As a result, vital processes such as birth, death, growth, metabolism, resource consumption, etc., vary among individuals. Birth rates of younger individuals are generally quite different from those of older individuals, mortality rates of larger individuals are usually different from those of smaller individuals, and so on. These differences can be considerable, with variances sometimes being larger within a population than between different populations.

The vital rates of individuals ultimately determine the dynamics of the entire population and how those dynamics are affected by the physical and biological environment. Accurate models of population level dynamics therefore require a connection to individual level vital rates. One such connection is provided by so-called "structured" population models.

The structured models considered in this monograph describe the distribution of individuals throughout different classes or categories. The definition of these classes is based upon individual differences that are important with regard to individual vital rates. For example, the categorization of individuals can be based upon chronological age, a measure of body size, life cycle stages, gender or genetic differences, biochemical makeup, spatial location, behavioral activities, etc. A structured model describes how individuals move in time among the defined classes. The model thereby describes the dynamics of the population class distribution and as a result the dynamics of the population as a whole.

To cite just a few examples, structured models are required for the study of questions dealing with the effects of maturation and gestation delays; intraclass competition (between, say, small and large individuals or between juveniles and adults); intraclass predation (cannibalism); juvenile bottlenecks (in which the individuals are subjected to heavy competition or predation before reaching reproductive maturity); selective predation on prey of certain ages or sizes; parasitization on specific life cycle stages of hosts; the relationship between body size and interspecific competitive success; mixed types of interactions (in which, for example, two species compete during one life cycle stage, but do not compete or even bear an entirely different relationship, such as a predator–prey relationship, at a different life cycle stage). See [55], [154], [323], [408], [435] for these and many other examples.

Structured models have many advantages. By making a link between the individual level and population level, they can account for dynamical behavior that unstructured models cannot. Environmental influences are very likely to affect different individuals differently. Therefore a structured model can more accurately describe and predict the importance that specific environmental factors have on the population's dynamics, as well as the consequences of changes in these factors. For example, an individual's movement through the structuring classes can cause delays in response to environmental changes, which can have a profound effect on the dynamics of the population as a whole. Another advantage of structured models is that they are more likely than unstructured models to involve parameters with clear biological interpretations that are amenable to measurement and thereby provide a greater opportunity for connection with data.

Many different types of mathematical equations have been used to formulate structured population models. One broad distinction between types of models is whether the variables are discrete or continuous. For example, model time can be continuous or it can be a discrete sequence of census times; individual body size might be a continuous variable or it might be classified by discrete size intervals; and so on. At one extreme all structuring variables, all state variables, and time are discrete; at the other extreme they are all continuous. In Chapter 1 models of the former type are considered; models of the latter type are considered in Chapter 2. Structured models can also be of mixed types. For example, so-called compartmental models describe in continuous time the dynamics of discrete state variable classes. Examples of models discrete in time but continuous in the state variables can be found in [267], [268]. We will not consider such mixed types in this monograph.

Both discrete and continuous structured models have a long tradition of use. Both have their advantages and disadvantages. Discrete models, for example, are usually easy to construct from the life cycle history of the population. Generally, discrete models avoid many technical difficulties that continuous models entail (e.g., the difficulties surrounding partial differential equations concerning well posedness of initial value problems, numerical simulations, rigorous justification of linearization procedures, etc.). Indeed, discrete models with arbitrarily general structuring offer no particular difficulties. For continuous models of such generality, however, severe difficulties arise with regard to even the fundamental questions of existence and uniqueness of solutions. Due to such difficulties a complete and rigorous theory of continuous models has been worked out only for restricted types of structuring (e.g., age structure). Furthermore, by the recursive nature of the equations involved, discrete models are extraordinarily easy to simulate on computers. Also, stochastic versions of discrete models are generally easier to construct and analyze.

On the other hand, models with discrete time cannot account for the dynamics between its census times. Unless the structuring classes are approximately discrete in the biological population, the dynamics of a model with discrete classes might be sensitive to how the classes are defined and measured. Of course, in principle discrete models can be constructed that approximate a continuous classification arbitrarily closely (e.g., by shrinking the length of size or age classes), but the model will become large in size and a continuous model might be more tractable. (Interestingly, a complete and rigorous study of discrete models as approximations to continuous models, or vice versa, is yet to be made; see [409].) For more discussion of discrete and continuous models see [408, Chapter 1].

The focus in this monograph is on the asymptotic dynamics of deterministic models. Except for a brief appearance in section 1.4, stochastic models are not considered. A general modeling theory for structured population dynamics is presented. A general treatment of equilibria and stability is given from the point of view of bifurcation theory. Bifurcation theory is particularly appropriate in theoretical population dynamics since one of the fundamental expectations of a mathematical model is a description of the circumstances under which a population has certain kinds of asymptotic dynamics and how these dynamics are predicted to change if perturbations occur. Also included in this monograph is a selection of applications. These applications were chosen to illustrate both the mathematical theories and a selection of biological problems that have received attention in the literature.

Attention will be focused on structuring variables related to physiological characteristics. Structured models could also be constructed using classes based upon spatial location and/or inhomogeneities in the physical habitat. Even though spatial structure is extremely important in population dynamics and there is a great deal of literature on spatial diffusion, migration, patchiness, etc., models that include explicit spatial structure are not considered.

This monograph is restricted almost exclusively to autonomous models. However, much of the theory and mathematical results presented here have been extended to periodically forced model equations, which are appropriate for populations subjected to periodic oscillations in their vital rates and/or environments (e.g., due to seasonality). See section 1.4.3.

The topics covered reflect the particular interests of the author and are not meant to be comprehensive, either mathematically or biologically. For example, there is a large literature on modeling the dynamics of cell growth which is not touched on. This monograph focuses exclusively on population dynamics and ecological interactions. No topics are included from many related disciplines, e.g., epidemiology, genetics, evolutionary biology, renewable resource management, or bioeconomics.

Chapter 1 contains a treatment of discrete models in discrete time which allow for very general structuring of a population. A methodology is presented for studying basic equilibrium and stability questions for such models from the perspective of bifurcation theory. Nonequilibrium dynamics are also covered, insofar as they arise from equilibrium destabilization and local bifurcations. A similar tact is taken in Chapter 2 for continuous models. However, only age-structured (and some simpler size-structured) models are treated, since the mathematical theory of continuous models with more general structuring is difficult and is not as complete. In Chapter 3 there appear some special types of structured models, both discrete and continuous, for which dynamical equations at the population level can be uncoupled from those at the individual level. Some details of local stability theory for (not necessarily invertible) maps are given in Appendix A. Other mathematical details appear in Appendices B and C.

This monograph is an outgrowth of a series of lectures given at a National Science Foundation Regional Conference arranged by the Conference Board of Mathematical Sciences and held at North Carolina State University, Raleigh, during June of 1997. I would like to thank John Franke and Abdul-Aziz Yakubu for organizing the conference and inviting me to give these lectures. The tremendous success of the conference was in large part due to their efforts and those of their supporting staff. Also contributing to this success were the many participants at the conference, who came from a wide variety of disciplines and who provided stimulating discussions and contributing talks. Special thanks are also due to Jim Yorke, whose provocative and insightful lectures were certainly a highlight of the meeting.

I would like to make a special acknowledgment to my colleagues R. F. Costantino, Brian Dennis, and R. A. Desharnais who have shown me how exciting and fruitful interdisciplinary collaborations can be.

I am very grateful to the National Science Foundation for its generous support of my research over the years.

With the usual caveat that responsibility for all errors is mine, I thank Shandelle Henson, William Mueller, and Joseph Watkins for their aid in proofreading this manuscript.

J. M. Cushing
University of Arizona

CHAPTER 1

Discrete Models

This chapter deals with dynamical models for structured populations in which both time and the structuring variables are discrete. These models involve systems of difference equations (or maps) of a type called *matrix population models* [55]. Matrix models are introduced in section 1.1. The asymptotic dynamics of autonomous matrix models are studied in section 1.2 from the point of view of bifurcation theory. Several applications are given in sections 1.3 and 1.4. These sections deal only with a single population. Matrix models for the interaction of several structured species are considered in section 1.5.

1.1 Matrix models

Suppose that the individuals of a population are categorized into a finite number of classes (e.g., by chronological age or some measure of body size). Let $x_i(t)$, for $i = 1, 2, \dots, m$, denote the number or density[1] of individuals in the ith class at time $t = 0, 1, 2, \dots$. Let

$$t_{ij} \doteq \begin{cases} \text{the fraction of } j\text{-class individuals expected to} \\ \text{survive and move to class } i \text{ per unit of time.} \end{cases}$$

Then at time $t+1$ the density of individuals in class i who were alive at time t is expected to be

$$\sum_{j=1}^{m} t_{ij} x_j(t).$$

Let

$$f_{ij} \doteq \begin{cases} \text{the expected number of (surviving) } i\text{-class} \\ \text{offspring per } j\text{-class individual per unit of time.} \end{cases}$$

Then at time $t+1$ the number of i-class newborns is expected to be

$$\sum_{j=1}^{m} f_{ij} x_j(t).$$

[1]Biomass, dry weight, or some other measure of population abundance of the individuals in the structuring classes could also be used.

If only birth and death processes are allowed, we have

$$x_i(t+1) = \sum_{j=1}^{m} t_{ij} x_j(t) + \sum_{j=1}^{m} f_{ij} x_j(t).$$

Using matrix notation these equations can be written in the compact form

$$x(t+1) = Px(t), \quad t = 0,\ 1,\ 2,\ 3, \dots, \tag{1.1}$$

where

$$x(t) \doteq \begin{bmatrix} x_1(t) \\ x_2(t) \\ \vdots \\ x_m(t) \end{bmatrix}$$

is the *class distribution vector* at time t and the matrix

$$P \doteq T + F$$

is the sum of the *transition matrix*

$$T \doteq [t_{ij}], \quad 0 \le t_{ij} \le 1, \quad \sum_{i=1}^{m} t_{ij} \le 1 \quad \text{for all } j \tag{1.2}$$

and the *fertility matrix*

$$F \doteq [f_{ij}], \quad f_{ij} \ge 0. \tag{1.3}$$

The quantities f_{ij} and t_{ij} are built from "submodels" based on class specific hypotheses about these vital quantities. These submodels would take into consideration relevant class specific rates of mortality, fertility, resource availability and consumption, metabolism, body growth, etc. These rates, and hence t_{ij} and f_{ij}, might be related to population crowding or so-called *density effects* (e.g., due to competition), in which case they become functions of one or more of the class densities x_i. Recursion equations of the form (1.1) are called *matrix equations.* (Rates not proportional to class densities are not included in this model. For example, immigration and emigration rates might be of such a type.)

The matrix P is called the *projection matrix.*[2] If P is constant, then the matrix equation (1.1) is *linear* and *autonomous.* If $P = P(t)$ depends explicitly on time t, then (1.1) is *linear* and *nonautonomous.* If $P = P(t, x(t))$ depends on $x(t)$, then (1.1) is *nonlinear.* In any case, given an initial class distribution $x(0)$, the recursion formula (1.1) defines a unique sequence $x(t)$, $t = 0, 1, 2, \dots$, called a *solution* (or more precisely a *forward solution*) of (1.1).[3] Since all entries in P are nonnegative it follows that $x_i(0) \ge 0$ implies that $x_i(t) \ge 0$ for all $t = 0, 1, 2, \dots$.

[2] P is not a projection matrix in the geometric or functional analysis sense.

[3] In general, P is not invertible and hence unique "backward" solutions are not defined. See A.1.

1.1.1 Notation and preliminaries. The set of real numbers and the set of nonnegative real numbers are denoted by R^1 and R^1_+, respectively. Let $R^m \doteq R^1 \times \cdots \times R^1$ and $R^m_+ \doteq R^1_+ \times \cdots \times R^1_+$ denote m-fold Cartesian products. For $x = [x_i] \in R^m$ we define the vector norm

$$|x| \doteq \sum_{i=1}^{m} |x_i| .$$

The transpose of x will be denoted by x^τ. Then the usual inner product is $x^\tau y$. The boundary of R^m_+ is denoted by ∂R^m_+. For a matrix $P = [p_{ij}]$ we use the operator norm

$$\|P\| \doteq \sup_{|x|\neq 0} \frac{|Px|}{|x|} = \sup_{|x|=1} |Px| .$$

Then $|Px| \leq \|P\| \, |x|$. Let $I(q_1, q_2)$ denote the set of integers i satisfying $q_1 < i < q_2$ (q_1 can be $-\infty$ and/or q_2 can be $+\infty$). Let $I[q_1, q_2) = I(q_1, q_2) \cup \{q_1\}$, etc.

By a *positive (nonnegative)* vector $x > 0$ ($x \geq 0$) or matrix $P > 0$ ($P \geq 0$) we mean all entries are positive (nonnegative). *Eigenvalues* r (real or complex) and *right and left eigenvectors* v and w (real or complex) of a matrix M satisfy $Mv = rv$, $v \neq 0$, and $w^\tau M = rw^\tau$, $w \neq 0$, respectively. A positive, *strictly dominant* eigenvalue r_1 satisfies $r_1 > |r|$ for all other eigenvalues r. The *algebraic multiplicity* of an eigenvalue r is the multiplicity of r as a root of the characteristic polynomial $\det(rI - M)$. An eigenvalue is *algebraically simple* if its algebraic multiplicity is equal to 1. In this case, all right and left eigenvectors have the form cv and cw, $c \in R^1$. The *characteristic values* of a matrix M are the reciprocals of the nonzero real eigenvalues of M.

Of particular interest in population models are nonnegative matrices $P \geq 0$. A remarkable theorem of Perron states that a positive matrix necessarily has a positive, (algebraically) simple, strictly dominant eigenvalue and to this eigenvalue there correspond positive right and left eigenvectors [343]. Matrices of interest in population dynamics are not always positive, however. A famous theorem of Frobenius adapts Perron's theorem to nonnegative matrices under the assumption of irreducibility [176], [179]. A matrix is *reducible* if a permutation of its rows and corresponding columns (which amounts to the reordering of the classes in a structured model) results in a block triangular matrix, i.e., a matrix in the form

$$\begin{bmatrix} A & 0 \\ B & C \end{bmatrix} .$$

If no such permutation exists, the matrix is *irreducible*. Equivalently, a matrix is irreducible if and only if its associated graph is strongly connected. The Frobenius theorem states that a nonnegative, irreducible matrix has a positive, (algebraically) simple eigenvalue whose magnitude is exceeded by no other eigenvalue and to this eigenvalue there correspond *positive* right and left eigenvectors. If, in addition, this maximal eigenvalue is strictly dominant, then the matrix is called *primitive*. It turns out that under these conditions there cannot exist two independent nonnegative eigenvectors [179, p. 63]. A necessary and sufficient condition that a nonnegative matrix $P \geq 0$ be irreducible and primitive is that there exists a positive integer j such that $P^j > 0$. See [179] for proofs of these facts.

THEOREM 1.1.1. *(Perron/Frobenius). If a nonnegative matrix is irreducible and primitive, then it has a positive, (algebraically) simple, strictly dominant eigenvalue and to this eigenvalue there correspond positive right and left eigenvectors. Moreover, there cannot exist two independent nonnegative right (or left) eigenvectors.*

1.1.2 Linear models. Consider the linear autonomous matrix equation (1.1) when $P \geq 0$ is a constant projection matrix. Matrix equations of this type were first used to describe the dynamics of age-structured populations [281], [282], [288]. In these so-called *Leslie matrix models*, a population is divided into age (since birth) categories, all of which have the same length as the discrete census time. For this reason the transition matrix T has the form

$$T = \begin{bmatrix} 0 & 0 & \cdots & 0 & 0 \\ t_{21} & 0 & \cdots & 0 & 0 \\ 0 & t_{32} & \cdots & 0 & 0 \\ \vdots & \vdots & \ddots & \vdots & \vdots \\ 0 & 0 & \cdots & t_{m,m-1} & 0 \end{bmatrix}, \quad 0 < t_{i,i-1} \leq 1.$$

Since newborns necessarily lie in the first age class, the fertility matrix has the form

$$F = \begin{bmatrix} f_{11} & f_{12} & \cdots & f_{1,m-1} & f_{1m} \\ 0 & 0 & \cdots & 0 & 0 \\ 0 & 0 & \cdots & 0 & 0 \\ \vdots & \vdots & \ddots & \vdots & \vdots \\ 0 & 0 & \cdots & 0 & 0 \end{bmatrix}, \quad 0 \leq f_{ij}$$

and the projection matrix

$$P = \begin{bmatrix} f_{11} & f_{12} & \cdots & f_{1,m-1} & f_{1m} \\ t_{21} & 0 & \cdots & 0 & 0 \\ 0 & t_{32} & \cdots & 0 & 0 \\ \vdots & \vdots & \ddots & \vdots & \vdots \\ 0 & 0 & \cdots & t_{m,m-1} & 0 \end{bmatrix} \tag{1.4}$$

is a so-called *Leslie matrix*. If there are k juvenile classes, then $f_{11} = f_{12} = \cdots = f_{1k} = 0$. If $f_{1m} > 0$, then P is irreducible. When is P primitive? Let $f_{1,m_1}, f_{1,m_2}, \ldots, f_{1,m_j}, f_{1m}$, $m_j \neq m$, be the nonzero fertilities appearing in the first row of F listed in order. The Leslie matrix P is primitive if and only if the greatest common divisor of the integers $m - m_j, m_j - m_{j-1}, \ldots, m_2 - m_1$ is equal to 1 [179], [245]. In particular, P is primitive if there are two consecutive fertile age categories.

A mathematical generalization of the Leslie matrix is given by the transition matrix

$$T = \begin{bmatrix} t_{11} & 0 & \cdots & 0 & 0 \\ t_{21} & t_{22} & \cdots & 0 & 0 \\ 0 & t_{32} & \cdots & 0 & 0 \\ \vdots & \vdots & \ddots & \vdots & \vdots \\ 0 & 0 & \cdots & t_{m,m-1} & t_{mm} \end{bmatrix},$$

$$0 < t_{i,i-1} \leq 1, \quad 0 \leq t_{ii} \leq 1.$$

This type of transition matrix has been utilized for populations whose individuals are categorized by (increasing) size classes. The diagonal entry t_{ii} is the fraction of individuals in size class i who survive and remain in class i after one time unit, and $t_{i+1,i}$ is the fraction that survives and moves to the next largest size category $i+1$. No individual can shrink in size or grow more than one class in one unit of time. If all newborns lie in the smallest size class, the resulting projection matrix

$$P = \begin{bmatrix} t_{11}+f_{11} & f_{12} & \cdots & f_{1,m-1} & f_{1m} \\ t_{21} & t_{22} & \cdots & 0 & 0 \\ 0 & t_{32} & \cdots & 0 & 0 \\ \vdots & \vdots & \ddots & \vdots & \vdots \\ 0 & 0 & \cdots & t_{m,m-1} & t_{mm} \end{bmatrix} \tag{1.5}$$

is called an *Usher matrix* [410], [411], [412] (or the *standard size-classified matrix* [55]). An Usher matrix is irreducible if all $t_{i,i-1} > 0$ and $f_{1m} > 0$. It is also primitive if two successive size classes are fertile.

Most projection matrices used in applications are irreducible and primitive. An example of an irreducible Leslie matrix that is not primitive is the juvenile/adult model

$$P = \begin{bmatrix} 0 & f_{12} \\ t_{21} & 0 \end{bmatrix}.$$

If adults are allowed to live more than one unit of time, the resulting modified Leslie matrix

$$P = \begin{bmatrix} 0 & f_{12} \\ t_{21} & t_{22} \end{bmatrix}, \quad t_{22} > 0,$$

is irreducible and primitive.

Consider the general linear autonomous matrix equation

$$x(t+1) = Px(t), \quad x(0) \in R_+^m, \tag{1.6}$$

and its solution

$$x(t) = P^t x(0), \quad t \in I[0, +\infty).$$

If P is nonnegative, irreducible, and primitive, then by Theorem 1.1.1 there is a simple, positive, and strictly dominant eigenvalue r and positive right and left eigenvectors $v > 0$ and $w > 0$. For simplicity, suppose P is diagonalizable and let $v_1, v_2, \ldots, v_m$ be a basis of right eigenvectors and $w_1, w_2, \ldots, w_m$ a basis of left eigenvectors, where $v_1 = v$ and $w_1 = w$. Write,

$$x(0) = cv + \sum_{i=2}^{m} c_i v_i.$$

Since the inner products $w^\tau v_i = 0$ for $i \neq 1$, it follows that $c = w^\tau x(0)/w^\tau v \geq 0$. If $x(0) \neq 0$, then $c > 0$. Let r_i, $i \in I[2, m]$, denote the remaining $m-1$ eigenvalues of P. Let

$$p(t) \doteq |x(t)| = \sum_{i=1}^{m} x_i(t)$$

denote the *total population size*. From

$$\frac{x(t)}{p(t)} = \frac{cr^t v + \sum_{i=2}^m c_i r_i^t v_i}{|cr^t v + \sum_{i=2}^m c_i r_i^t v_i|} = \frac{cv + \sum_{i=2}^m c_i \left(\frac{r_i}{r}\right)^t v_i}{\left|cv + \sum_{i=2}^m c_i \left(\frac{r_i}{r}\right)^t v_i\right|}$$

and $r > |r_i|$ for all $i \in I[2, m]$, we find that

$$\lim_{t \to +\infty} \frac{x(t)}{p(t)} = \frac{v}{|v|}. \tag{1.7}$$

The unit eigenvector $v/|v|$ is called the *stable distribution*, and the limit (1.7) is called the *(strong) ergodic property* [55], [64], [65], [66], [67], [245]. It remains true for nondiagonalizable projection matrices P as well (see [245] for a proof).

The limit (1.7) describes the asymptotic behavior of the *normalized distribution* $x(t)/p(t)$, regardless of the asymptotic dynamics of the class distribution $x(t)$ itself or that of the total population size $p(t)$. The equation

$$p(t+1) = \left| P \frac{x(t)}{p(t)} \right| p(t) \tag{1.8}$$

for the dynamics of $p(t)$ can be obtained directly from $p(t+1) = |Px(t)|$. This is a scalar, nonautonomous difference equation for $p(t)$ that is coupled with the equation (1.6) for $x(t)$. However, by (1.7) the coefficient of $p(t)$ in (1.8) satisfies

$$\lim_{t \to +\infty} \left| P \frac{x(t)}{p(t)} \right| = r.$$

This suggests that the asymptotic dynamics of $p(t)$ are related to those of the so-called *limiting equation*

$$q(t+1) = rq(t), \tag{1.9}$$

a particularly simple scalar recursion formula. The solution $q(t) = r^t q(0)$ tends, of course, to 0 if $r < 1$ and grows exponentially if $r > 1$. Is this true of the total population size $p(t)$ satisfying (1.8)? The answer is "yes," a fact proved in Appendix C. We summarize these results in the following theorem. (For further theorems of this type see [55], [64], [65], [66], [67].)

THEOREM 1.1.2. *(The Fundamental Theorem of Demography). Suppose that the nonnegative matrix* $P \geq 0$ *is irreducible and primitive. Let* r *be the strictly dominant eigenvalue of* P *and* $v > 0$ *be an associated eigenvector. Let* $x(t)$ *be the solution of the linear matrix equation* $x(t+1) = Px(t)$, $t \in I[0, +\infty)$, *with an initial state satisfying* $0 \leq x(0) \neq 0$, *and let* $p(t) = |x(t)|$. *Then*

(a) (1.7) *holds and*

(b) $\lim_{t\to+\infty} p(t) = 0$ *if* $r < 1$ *and* $\lim_{t\to+\infty} p(t) = +\infty$ *if* $r > 1$.

The limiting equation (1.9) is an example of a dynamical equation for a population level statistic derived from a structured model. The individual level parameters in the projection matrix are encapsulated in its dominant eigenvalue r and can therefore be related to population level dynamics.

The dominant eigenvalue r is the *(inherent) growth rate* of the population. The magnitude of r determines whether a population with positive initial size $p(0) > 0$ goes extinct ($r < 1$) or grows exponentially without bound ($r > 1$). Thus, a *bifurcation* occurs as r passes through the critical value 1. When $r = 1$ (and for no other value of r), there exist equilibrium class distributions, namely, any scalar multiple of the eigenvector v. The *extinction equilibrium* $x(t) = 0$ (or $p(t) = 0$) always exists for the matrix equation $x(t+1) = Px(t)$. If equilibria are viewed as functions of r, we see that there is an intersection (or bifurcation) of two equilibrium branches, the extinction equilibria and the nonzero (eigenvector) equilibria, which occurs at the *bifurcation point* $(r, x) = (1, 0)$ (or $(r, p) = (1, 0)$). This bifurcation is *vertical* because the *spectrum* for the nonextinction equilibria consists of the single point $r = 1$. See Fig. 1.1.

Another important quantity is the *inherent net reproductive number* n (sometimes called the net reproductive *value* or *rate* [184]). This quantity is defined to be the expected number of offspring per individual per lifetime. For a Leslie model a formula for n is obtained by summing the products of the expected number of offspring f_{1i} from each age class and the probability $t_{21}t_{32}\cdots t_{i,i-1}$ of reaching that age class. Thus,

$$n = \sum_{i=1}^{m} f_{1i} \prod_{j=1}^{i} t_{j,j-1}, \tag{1.10}$$

where for notational convenience t_{10} is defined to be 1. For a general projection matrix $P = T + F$ the inherent net reproductive number n can be defined as follows.

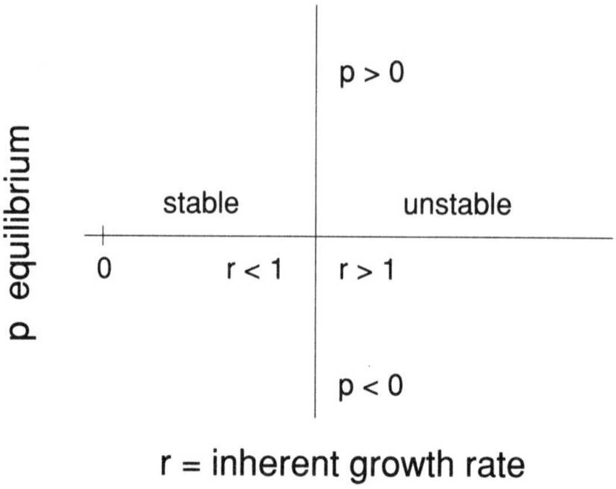

FIG. 1.1. *A plot of the inherent growth rate r against the total population size $p = |x|$ of the equilibria x of the linear matrix equation* (1.6) *shows two branches intersecting at $(r, p) = (1, 0)$. Namely, the branch of trivial equilibrium $(r, 0)$ intersects the branch of nontrivial equilibria $(1, p)$, $p \in R^1$. The extinction equilibrium $p = 0$ is stable for $r < 1$ and is unstable for $r > 1$.*

DEFINITION 1. *Let T and F be the transition and fertility matrices* (1.2) *and* (1.3). *If $I-T$ is invertible and $F(I-T)^{-1}$ has a positive, (algebraically) simple, strictly dominant eigenvalue n and a nonnegative eigenvector $u \geq 0$, then n is called the* inherent net reproductive number *for the projection matrix $P = T+F$.*

Necessary for the invertibility of $I-T$ is that at least one of the column sums $\sum_{i=1}^{m} t_{ij}$ of T is strictly less than 1. A sufficient condition is that all column sums be strictly less than 1, i.e., $\sum_{i=1}^{m} t_{ij} < 1$ for each $j \in [1, m]$. This biologically reasonable condition means that over each time interval there is some loss to every class of individuals (e.g., due to deaths). Since $(I-T)^{-1} = \sum_{i=0}^{\infty} T^i \geq 0$, we see that $F(I-T)^{-1}$ is a nonnegative matrix. Sufficient, but not necessary, for the definition of n is that $F(I-T)^{-1}$ be irreducible and primitive.

Definition 1 is a mathematical one. What is the biological interpretation of n? By Definition 1, n is given by [179]

$$n = \max_{x \in R_+^m/\{0\}} \frac{|F(I-T)^{-1}x|}{|x|}.$$

From this formula we can obtain a biological interpretation for n as follows. The i, j entry in $(I-T)^{-1} = I + T + T^2 + \cdots$ is the expected amount of time that an individual starting in class j will spend in class i over the course of its lifetime. The i, j entry in the matrix $F(I-T)^{-1}$ is the expected number of i class offspring that an individual born into class j will produce over the course of its lifetime. Thus, the total expected offspring of an individual born in class j over its lifetime is the sum of elements in the jth column of $F(I-T)^{-1}$.

Consider first the simplest and most common case in matrix models in which there is only one newborn class. Then F has only one nonzero row which we take, without loss of generality, to be the first row, and consequently only the first row of $F(I-T)^{-1}$ is nonzero. This means, in this case, that $F(I-T)^{-1}$ has 0 as an $m-1$ repeated eigenvalue and the dominant eigenvalue (the inherent net reproductive number, by definition) is the first row, first column entry. Since this entry is the only nonzero entry in the first column, it is equal to the sum of entries in the first column. It follows that *for the case of a single newborn class the inherent net reproductive number n is the expected number of offspring per newborn over the course of its lifetime.*

More generally, suppose that there are $j > 1$ newborn classes, which without loss of generality we list first so that the last $m - j$ rows of F consist of zeros only. If we denote the ith *row* of F by f_i^T and the jth *column* of $(I - T)^{-1}$ by e_j, i.e.,

$$F = \begin{bmatrix} f_1^T \\ \vdots \\ f_m^T \end{bmatrix}, \qquad (I-T)^{-1} = \begin{bmatrix} e_1 & \cdots & e_m \end{bmatrix},$$

then

$$F(I-T)^{-1} = \begin{bmatrix} f_1^T e_1 & \cdots & f_1^T e_j & f_1^T e_{j+1} & \cdots & f_1^T e_m \\ \vdots & \ddots & \vdots & \vdots & \ddots & \vdots \\ f_j^T e_1 & \cdots & f_j^T e_j & f_j^T e_{j+1} & \cdots & f_j^T e_m \\ 0 & \cdots & 0 & 0 & \cdots & 0 \\ \vdots & \ddots & \vdots & \vdots & \ddots & \vdots \\ 0 & \cdots & 0 & 0 & \cdots & 0 \end{bmatrix}. \tag{1.11}$$

Thus, 0 is an eigenvalue of multiplicity at least $m-j$. The dominant eigenvalue of $F(I-T)^{-1}$ is the dominant eigenvalue of the submatrix S_j from the upper left-hand corner; thus,

$$n = \max_{y \in R_+^j/\{0\}} \frac{|S_j y|}{|y|}. \tag{1.12}$$

For a class distribution of individuals $x \in R_+^m$ the expected distribution of offspring over the course of their lifetimes is $F(I-T)^{-1}x$. If we consider the expected distribution of offspring from a group of newborn members of the population, then x has nonzero entries in only its first j entries so that

$$x = \begin{bmatrix} y \\ 0 \end{bmatrix},$$

where $y \in R_+^j$ is the distribution of those newborns. Then the distribution of expected offspring from this group is

$$F(I-T)^{-1}x = \begin{bmatrix} S_j y \\ 0 \end{bmatrix}$$

and the total expected offspring is $\left|F(I-T)^{-1}x\right| = |S_j y|$. From (1.12) we see that n *is the maximum expected per capita lifetime number of offspring from newborn individuals, where the maximum is taken over all possible newborn class distributions.* This maximum is attained by class distributions proportional to the eigenvector u associated with n as the dominant eigenvalue of $F(I-T)^{-1}$.

Note that any eigenvalue of $F(I-T)^{-1}$ is also an eigenvalue of $(I-T)^{-1}F$ and vice versa, since both matrices have the same characteristic polynomial. Thus, the inherent net reproductive number n is also a strictly dominant positive eigenvalue of the matrix $(I-T)^{-1}F$. It is easy to show that $(I-T)^{-1}u \geq 0$ is a right eigenvector of $(I-T)^{-1}F$ associated with n. The following theorem is proved in [115, Theorem 3 and Corollary 7].

THEOREM 1.1.3. *Consider a nonnegative projection matrix* $0 \leq P = T + F$, *where the transition and fertility matrices* T *and* F *satisfy* (1.2)–(1.3). *Assume that* P *has a positive, (algebraically) simple, strictly dominant eigenvalue* r *with a positive right eigenvector* $v > 0$. *Assume further that* T *and* F *satisfy the requirements of Definition* 1 *with* $(I-T)^{-1}u > 0$.

Then $r < 1$ *if and only if* $n < 1$ *and* $r > 1$ *if and only if* $n > 1$. *(Thus,* $r = 1$ *if and only if* $n = 1$.*)*

Moreover, $n < 1$ *implies that* $n \leq r < 1$ *and* $n > 1$ *implies that* $1 < r \leq n$.

Thus, a population grows exponentially if its inherent net reproductive number is $n > 1$ and dies exponentially if $n < 1$ (not an unexpected result given the biological interpretation of n).

Whereas formulas for the population growth rate r in terms of the entries in the projection matrix P are not in general available, such formulas for n are available for broad classes of projection matrices [115]. For example, in the common case of models with a single newborn class we have from (1.12) that the inherent net reproductive number is given by the inner product of the first (fertility) row of F with the first column of $(I-T)^{-1}$, namely,

$$n = f_1^t e_1. \tag{1.13}$$

A formula for n is thus available when a formula for $(I-T)^{-1}$ is available.

As an example, consider an irreducible and primitive Usher matrix (1.5) with $0 \leq t_{jj} < 1$. Then

$$(I-T)^{-1} = \begin{bmatrix} \frac{1}{1-t_{11}} & 0 & \cdots & 0 \\ \frac{t_{21}}{(1-t_{11})(1-t_{22})} & \frac{1}{1-t_{22}} & \cdots & 0 \\ \vdots & \vdots & \cdots & 0 \\ \prod_{j=1}^{i} \frac{t_{j,j-1}}{1-t_{jj}} & \prod_{j=2}^{i} \frac{t_{j,j-1}}{1-t_{jj}} & \cdots & 0 \\ \vdots & \vdots & \ddots & \vdots \\ \prod_{j=1}^{m} \frac{t_{j,j-1}}{1-t_{jj}} & \prod_{j=2}^{m} \frac{t_{j,j-1}}{1-t_{jj}} & \cdots & \frac{1}{1-t_{mm}} \end{bmatrix}. \tag{1.14}$$

The matrix

$$F(I-T)^{-1} = \begin{bmatrix} \sum_{i=1}^m f_{1i} \prod_{j=1}^i \frac{t_{j,j-1}}{1-t_{jj}} & \cdots & f_{1m}\frac{1}{1-t_{mm}} \\ 0 & \cdots & 0 \\ \vdots & \ddots & \vdots \\ 0 & \cdots & 0 \end{bmatrix}$$

has dominant eigenvalue

$$n = \sum_{i=1}^m f_{1i} \prod_{j=1}^i \frac{t_{j,j-1}}{1-t_{jj}} \tag{1.15}$$

(where for notational convenience we have defined $t_{10} = 1$) with a unit eigenvector

$$u = \begin{bmatrix} 1 \\ 0 \\ \vdots \\ 0 \end{bmatrix} \geq 0.$$

Also,

$$(I-T)^{-1}u = \begin{bmatrix} \frac{1}{1-t_{11}} \\ \frac{t_{21}}{(1-t_{11})(1-t_{22})} \\ \frac{t_{21}t_{32}}{(1-t_{11})(1-t_{22})(1-t_{33})} \\ \vdots \\ \frac{t_{21}t_{32}t_{43}\cdots t_{m,m-1}}{(1-t_{11})(1-t_{22})\cdots(1-t_{mm})} \end{bmatrix} > 0.$$

Thus, Theorem 1.1.3 applies.

As a second example, consider an Usher-type model with two classes of newborns (taken without loss of generality to be classes $i = 1$ and 2) in which only the largest class is fertile. For this case the fertility matrix is

$$F = \begin{bmatrix} 0 & 0 & \cdots & 0 & f_{1m} \\ 0 & 0 & \cdots & 0 & f_{2m} \\ 0 & 0 & \cdots & 0 & 0 \\ \vdots & \vdots & \ddots & \vdots & \vdots \\ 0 & 0 & \cdots & 0 & 0 \end{bmatrix}, \quad f_{1m} > 0, \quad f_{2m} > 0,$$

and $(I-T)^{-1}$ is again given by (1.14). From (1.11) we obtain the formula

$$n = (f_{1m} + f_{2m}(1-t_{11})) \prod_{j=1}^m \frac{t_{j,j-1}}{1-t_{jj}} > 0$$

for the inherent net reproductive number. Also,

$$u = \begin{bmatrix} f_{1m} \\ f_{2m} \\ 0 \\ \vdots \\ 0 \end{bmatrix} \geq 0$$

and

$$(I-T)^{-1}u = \begin{bmatrix} f_{1m}\frac{1}{1-t_{11}} \\ f_{1m}\frac{t_{21}}{(1-t_{11})(1-t_{22})} + f_{2m}\frac{1}{1-t_{22}} \\ \vdots \\ f_{1m}\prod_{j=1}^{i}\frac{t_{j,j-1}}{1-t_{jj}} + f_{2m}\prod_{j=2}^{i}\frac{t_{j,j-1}}{1-t_{jj}} \\ \vdots \\ f_{1m}\prod_{j=1}^{m}\frac{t_{j,j-1}}{1-t_{jj}} + f_{2m}\prod_{j=2}^{m}\frac{t_{j,j-1}}{1-t_{jj}} \end{bmatrix} > 0$$

so that Theorem 1.1.3 applies.

1.1.3 Nonlinear models. Linear matrix models (1.1) with constant projection matrices P imply exponential dynamics. In the case of exponential growth, linear models therefore cannot describe the long term (asymptotic) dynamics of populations. It follows that models suitable for asymptotic dynamics and population growth must involve projection matrices P whose entries (namely, the fertility and transition rates appearing in F and T) are not constant in time. If P depends implicitly on time through a dependence on class densities (called "density effects"), then the matrix model (1.1) is nonlinear. A projection matrix may also depend explicitly on time t (deterministically and/or stochastically).

One way that nonlinear projection matrices arise is through fractional decreases in one or more entries in the transition matrix T and/or the fertility matrix F that are dependent upon class densities. For example, suppose the transition probability t_{ij} of a j class individual to class i is adversely affected in some way by an encounter with (or simply the presence of) a k class individual. If the probability of such an encounter is taken to be (approximately) proportional to the length of time involved, then during an interval Δt of time the probability of no encounter occurring between a j class individual and a k class individual is approximately $1 - \omega_{ij,k}\Delta t$, where $\omega_{ij,k}$ is the constant of proportionality. If there are $x_k(t)$ individuals of class k present at time t and the encounters are independent, then the probability of no encounter during an interval Δt of time is approximately $(1 - \omega_{ij,k}\Delta t)^{x_k}$. During one full unit of time the probability of no encounter is (assuming independence)

$$\pi(\Delta t) = \prod_{l=1}^{1/\Delta t} (1 - \omega_{ij,k}\Delta t)^{x_k(t+(l-1)\Delta t)}.$$

Then

$$\begin{aligned} \ln \pi(\Delta t) &= \ln(1-\omega_{ij,k}\Delta t)\sum_{l=1}^{1/\Delta t} x_k(t+(l-1))\Delta t \\ &\approx -\omega_{ij,k}\sum_{l=1}^{1/\Delta t}(x_k(t+(l-1)\Delta t))\,\Delta t. \end{aligned}$$

Letting $\Delta t \to 0$ we obtain

$$\lim_{\Delta t\to 0}\pi(\Delta t) = \exp\left(-\omega_{ij,k}\int_t^{t+1} x_k(s)ds\right).$$

If we approximate $\int_t^{t+1} x_k(s)ds \approx x_k(t)$, then the probability of no encounters with k class individuals during one unit of time is approximately equal to $\exp(-\omega_{ij,k}x_k(t))$. Finally, if π_{ij} is the probability that a j class individual survives one unit of time and moves to class i independent of the presence of k class individuals, then

$$t_{ij} = \pi_{ij}\exp(-\omega_{ij,k}x_k(t))$$

is the i, j entry in the transition matrix T at time t. If other classes also affect t_{ij} (independently), then additional exponential factors appear and one obtains

$$t_{ij} = \pi_{ij}\exp(-p_{ij}(t)),$$

where

$$p_{ij}(t) \doteq \sum_{k=1}^{m}\omega_{ij,k}x_k(t)$$

is now a *weighted total population size.*

Exponential nonlinearities of the form $\exp(-cp_{ij})$ are often referred to as *Ricker-type* nonlinearities. (See [294], [391] for further treatments of exponential types of nonlinearities in age-structured models.) Other nonlinear expressions also appear frequently in applications. These may be derived from modeling assumptions concerning the class specific interactions between individuals or they may be ad hoc nonlinearities with desired qualitative features. Examples include the *Beverton–Holt* nonlinearity $(1+cp_{ij})^{-1}$ or, more generally, $(1+cp_{ij}^{\alpha})^{-1}$, and the modified exponential $\exp(-cp_{ij}^{\alpha})$, $\alpha > 0$. In general, if population density effects are deleterious, then appropriate entries in T and F are multiplied by fractions which are *decreasing* functions of weighted total population sizes. Thus, $t_{ij} = \pi_{ij}\varphi_{ij}(p_{ij})$ and/or $f_{ij} = b_{ij}\psi_{ij}(p_{ij})$, where φ_{ij}, $\psi_{ij} \in C^1(R_+^1, [0,1])$ are decreasing functions of their arguments.

Examples of nonlinear projection matrices with various types of nonlinearities taken from the literature are

$$
\frac{1}{1+\alpha p}\begin{bmatrix} b_{11} & b_{12} & \cdots & b_{1,m-1} & b_{mm} \\ \pi_{21} & 0 & \cdots & 0 & 0 \\ 0 & \pi_{32} & \cdots & 0 & 0 \\ \vdots & \vdots & \ddots & \vdots & \vdots \\ 0 & 0 & \cdots & \pi_{m,m-1} & 0 \end{bmatrix}, \tag{1.16}
$$

$$
\begin{bmatrix} b_{11}\psi(p) & b_{12}\psi(p) & \cdots & b_{1,m-1}\psi(p) & b_{mm}\psi(p) \\ \pi_{21} & 0 & \cdots & 0 & 0 \\ 0 & \pi_{32} & \cdots & 0 & 0 \\ \vdots & \vdots & \ddots & \vdots & \vdots \\ 0 & 0 & \cdots & \pi_{m,m-1} & 0 \end{bmatrix}, \quad \psi'(p) < 1, \tag{1.17}
$$

with $p = |x|$, and

$$
\begin{bmatrix} 0 & b_1 & b_2 \\ \varphi(p) & 0 & 0 \\ 1-\varphi(p) & 0 & 0 \end{bmatrix}, \quad p = x_1 + \sum_{i=2}^{3} \omega_j x_j, \quad \varphi'(p) < 0, \tag{1.18}
$$

$$
\begin{bmatrix} 0 & 0 & b\exp(-c_{el}x_1 - c_{ea}x_3) \\ 1-\mu_l & 0 & 0 \\ 0 & \exp(-c_{pa}x_3) & 1-\mu_a \end{bmatrix}. \tag{1.19}
$$

The matrix (1.16) was introduced by Leslie [282]. It assumes that all transitions and fertilities are decreased by the same Beverton–Holt-type fraction $(1+cp)^{-1}$ as a function of total population size. In the matrix (1.17) only fecundity and/or newborn survival rates are affected by total population size p. This matrix is used in [283] with $\psi(p) = \exp(-cp)$ and with other types of nonlinearities in [160]. The matrix with two juvenile stages (1.19) was introduced in [132]; it utilizes two exponential nonlinearities involving two different weighted total population sizes. The matrix (1.18) is from a size-structured model with a juvenile stage and two adult classes [112]. In all of these examples, increased population density is deleterious in the sense that survival and/or fertility rates decrease with increased density.

Models in which increased population density is not deleterious, but is in fact advantageous (an effect called "strict depensation" or the "Allee effect"), are of biological interest [5], [6], [130]. Nonlinear factors that can be used to model this effect include

$$
x^\alpha \exp(-cx), \quad \frac{cx^\alpha}{1+cx^\alpha}, \quad 1-(1-cx)^\alpha, \quad c>0, \quad \alpha>1.
$$

An example is given by the projection matrix

$$\begin{aligned} P &= \begin{bmatrix} s_s\left(1 - M(x_1)\right) & \frac{1}{2} f s_j D(x_1, x_2) + p_f \\ 2 s_s M(x_1) & p_s \end{bmatrix}, \\ M(x_1) &\doteq 1 - \left(1 - 2\frac{x_1}{T}\right)^p, \\ D(x_1, x_2) &\doteq 1 - \left(1 - \frac{U - x_2 - x_1}{T}\right)^q \end{aligned}$$

from a model for spotted owl dynamics [209], [272], [275] in which the transition rate t_{21} has an Allee effect.

For other examples of matrix equation applications to population dynamics see [55], [71], [72], [73], [93], [99], [98], [101], [110], [112], [116], [151], [152], [153], [154], [160], [178], [238], [283], [294], [299], [335], [342], [376], [386], [391], [408], [410], [413], [415], [448] (and references cited therein).

Multispecies interactions among several species can be modeled by coupled systems of matrix models in which the projection matrices of some species are functions of the densities of other species. For example, the system

$$\begin{aligned} x(t+1) &= P_1\left(x(t), y(t)\right) x(t), \\ y(t+1) &= P_2\left(x(t), y(t)\right) y(t) \end{aligned}$$

denotes a two species interaction in which the entries of the transition and/or fertility matrices of each species are functions of the class densities of the other species. Predator–prey, competition, and mutualistic interactions can be modeled in this way. So can mixed-type interactions in which some stages might compete while others have a predator–prey relationship [435]. The vectors $x(t)$ and $y(t)$, and hence the projection matrices P_1 and P_2, need not be of the same size; i.e., the number of structure classes (or their type) need not be the same for both species.

Nonautonomous matrix equations

$$x(t+1) = P\left(t, x(t)\right) x(t)$$

model situations in which transition or fertility rates depend explicitly on time in some manner. For example, one or more of these rates might vary periodically according to some environmental periodicity (e.g., seasonal or daily fluctuations), or stochastically [74], [90], [107], [227], [232], [340].

1.2 Autonomous single species models

Consider the general nonlinear autonomous matrix equation

$$x(t+1) = P\left(x(t)\right) x(t), \quad t \in I[0, +\infty). \tag{1.20}$$

We will always assume that $P(x)$ is nonnegative for nonnegative x and is continuously differentiable in x. Specifically,

$$P(x) = \left[p_{ij}(x)\right], \quad p_{ij} \in C^1(\Omega, R^1_+), \tag{1.21}$$

where Ω is an open set in R^m that contains the (closed) nonnegative cone R_+^m. Clearly, a unique (forward) solution $x(t)$ is defined for each initial condition $x(0) \in \Omega$. Moreover, $x(0) \geq 0$ implies that $x(t) \geq 0$ for all $t \in I[0, +\infty)$; i.e., R_+^m is forward invariant. Thus, matrix models have no mathematical difficulties associated with the existence and uniqueness of solutions of initial value problems, nor with the positivity of solutions.

The study of the asymptotic dynamics associated with (1.20) naturally begins with a study of equilibrium solutions $x(t) = x \in R_+^m$ (i.e., time independent solutions). The equilibrium equation associated with (1.20) is the nonlinear algebraic system $x = P(x)x$. We are interested only in nonnegative solutions of this equation. Clearly, $x = 0$ is such a solution and it will be referred to as the *extinction equilibrium*. With regard to nonzero, nonnegative equilibria $0 \neq x \geq 0$, there are two basic problems: existence and stability.

An equilibrium $x \in R_+^m$ of equation (1.20) is called *(locally) stable* if for each $\varepsilon > 0$ there exists a $\delta = \delta(\varepsilon) > 0$ such that $|x(0) - x| < \delta$ implies that $|x(t) - x| < \varepsilon$ for all $t \in I[0, +\infty)$. If x is not stable, then it is called *unstable*. The equilibrium x is an *attractor* if there exists a $\delta > 0$ such that $|x(0) - x| < \delta$ implies that $\lim_{t \to +\infty} |x(t) - x| = 0$. If x is a stable attractor then it is called *(locally) asymptotically stable.*

The Jacobian of $P(x)x$ of the right-hand side of equation (1.20) is important with regard to the stability properties of an equilibrium x. This Jacobian is given by the formula[4]

$$J(x) = P(x) + [(\partial_i P)\, x]\,, \quad \partial_i \doteq \frac{\partial}{\partial x_i}.$$

A matrix is *hyperbolic* if all its eigenvalues ζ satisfy $|\zeta| \neq 1$. An equilibrium x is called hyperbolic if the Jacobian $J(x)$ is hyperbolic. The (local) stability properties of a hyperbolic equilibrium can be determined from the eigenvalues of $J(x)$. While this fact is well known for (local) diffeomorphisms [156], [193], [436] (i.e., if $\zeta = 0$ is not an eigenvalue of $J(x)$), the supporting theorems for a local stability analysis without this restriction are not. Therefore these theorems and their proofs are given in Appendix A.2. Among other things these theorems imply that an equilibrium x is (locally) asymptotically stable if all eigenvalues of the Jacobian $J(x)$ satisfy $|\zeta| < 1$ and is unstable if at least one eigenvalue satisfies $|\zeta| > 1$. First we turn our attention to the extinction equilibrium $x = 0$.

1.2.1 The extinction equilibrium.

For the extinction equilibrium $x = 0$ the Jacobian $J(0)$ equals the matrix $P(0)$, which we call the *inherent projection matrix.* The inherent projection matrix is the projection matrix when all density effects are ignored; therefore it governs the dynamics at low population levels.

By assumption (1.21), the inherent projection matrix $P(0) \geq 0$ is nonnegative. Assume that $P(0)$ is also irreducible and primitive. Then by Theorem 1.1.1 there is a positive strictly dominant simple eigenvalue $r > 0$ (with positive left and

[4] $[(\partial_i P)\, x]$ is the matrix whose i^{th} column is $(\partial_i P(x))x$.

right eigenvectors $w > 0$ and $v > 0$). If $r < 1$, then the extinction equilibrium $x = 0$ is (locally asymptotically) stable. This means, of course, that populations with sufficiently small initial total population sizes $p(0) = |x(0)|$ will go extinct. It may not be true, however, that all initial conditions lead to extinction.

Under additional conditions a stronger stability statement can be made when $r < 1$. The most common assumption in population models is that density effects are deleterious at all density levels. This means that the entries $t_{ij} = t_{ij}(x)$ in the transition matrix T are decreasing (or at least nonincreasing) functions of the components of x and in particular $0 \leq x$ implies that $0 \leq T(x) \leq T(0)$; the same is commonly true for the fertility matrix $F = F(x)$.

Under the assumption that the projection matrix $P(x)$ satisfies

$$0 \leq P(x) \leq P(0) \quad \text{for} \quad x \in R_+^m, \tag{1.22}$$

it follows that $0 \leq x \leq y$ implies that $0 \leq P(x)x \leq P(0)x \leq P(0)y$. Since $x(0) \in R_+^m$ implies that $x(t+1) \in R_+^m$ for all $t \in I[0, +\infty)$, we find that

$$0 \leq x(t+1) = P(x(t))x(t) \leq P(0)x(t), \quad t \in I[0, +\infty).$$

If we let $y(t)$ denote the solution of the initial value problem

$$y(t+1) = P(0)y(t), \quad y(0) = x(0), \quad t \in I[0, +\infty),$$

a straightforward induction shows that $0 \leq x(t) \leq y(t)$, $t \in I[0, +\infty)$. If $r < 1$, then $|y(t)|$ tends exponentially to 0 as $t \to +\infty$ (cf. Theorem 1.1.2) and consequently so does $p(t) = |x(t)|$. Thus, (1.22) and $r < 1$ imply *global* extinction. (This result can be found in [8].)

If $r > 1$, then the extinction equilibrium is unstable. From the definition of instability it does not necessarily follow that no population will go extinct. In fact, if $x = 0$ is hyperbolic, we know from Theorem A.2.2 (see Appendix A.2) that there may exist a stable manifold of initial conditions near $x = 0$ whose solutions tend to 0 as $t \to +\infty$. However, this stable manifold is tangent at $x = 0$ to the stable manifold of the linearization. Since the Jacobian is $P(0)$, the stable manifold of the linearization cannot intersect the cone R_+^m (except at $x = 0$). Therefore the stable manifold cannot locally intersect the cone R_+^m (except at $x = 0$).

If, in addition to $r > 1$, all solutions with $x(0) \geq 0$ are bounded, a stronger statement can often be made. An assumption that usually implies that solutions of population models are bounded is that both transition probabilities and fertilities decrease with increased population densities. In fact, a stronger condition usually holds for population models, namely,

(1.23) there exists a constant $c > 0$ so that for each $x(0) \in R_+^m$ there is an integer $t' \geq 0$ such that $p(t) \leq c$ for all $t \in I[t', +\infty)$.

With this property (1.20) is called *point dissipative* [211]. Furthermore, the map $M : R_+^m \to R_+^m$ defined by $M(x) \doteq A(x)x$ is continuous on the metric space

R_+^m, and therefore maps bounded sets to bounded (hence precompact) sets. In the jargon of [211], M is completely continuous. By Lemma 2.3.1 of [211] M is "asymptotically smooth," and it follows from Theorem 2.4.6 of [211] that there exists a global (connected) attractor in R_+^m.[5] Using this fact, it can be further proved that no initial total population size will lead to extinction in the sense of the following definition.

DEFINITION 2. *The equation* (1.20) *is* uniformly persistent with respect to $x = 0$ *if there exists a constant* $\eta > 0$ *such that* $x(0) \in R_+^m / \{0\}$ *implies that* $\liminf_{t\to+\infty} p(t) \geq \eta$.

The proof of the following theorem appears in Appendix C.

THEOREM 1.2.1. *Consider the matrix equation* (1.20)–(1.21). *Assume that the inherent projection matrix* $P(0)$ *is irreducible, primitive, and hyperbolic, and let* $r > 0$ *be its strictly dominant eigenvalue.*

(a) *If* $r < 1$, *then* $x = 0$ *is (locally asymptotically) stable. If, in addition,* (1.22) *holds, then* $\lim_{t\to+\infty} p(t) = 0$ *for all* $x(0) \in R_+^m$.

(b) *If* $r > 1$, *then* $x = 0$ *is unstable. If, in addition,* (1.20) *is point dissipative, i.e.,* (1.23) *holds, then* (1.20) *is uniformly persistent with respect to the extinction equilibrium* $x = 0$.

For a "weak persistence" result see [8].

The following theorem gives some conditions under which (1.20) is point dissipative.

THEOREM 1.2.2. *Consider the matrix equation* (1.20) *with projection matrix* $P(x) = [t_{ij}(x)] + [f_{ij}(x)]$. *Suppose that there exist constants* $\delta > 0$ *and* $f > 0$ *such that for all* $x \in R_+^m$ *and* $i, j \in I[1, m]$

$$0 \leq t_{ij}(x) \leq 1 \quad \textit{and} \quad \sum_{i=1}^{m} t_{ij}(x) \leq \delta < 1 \tag{1.24}$$

and

$$0 \leq f_{ij}(x) \quad \textit{and} \quad |F(x)x| \leq f. \tag{1.25}$$

Then (1.23) *holds; i.e.,* (1.20) *is point dissipative.*

Proof. Since $|T(x)x| \leq \delta\,|x|$, we have for each $t \in I[0, +\infty)$

$$|x(t+1)| \leq \delta\,|x(t)| + f,$$

which by induction implies that for $t \in I[1, +\infty)$

$$|x(t)| \leq \delta^t\,|x(0)| + f \sum_{i=0}^{t-1} \left|\delta^i\right| \leq \delta^t\,|x(0)| + \frac{f}{1-\delta}.$$

[5] A *global attractor* is an invariant set $\mathcal{A} \subseteq R_+^m$ (i.e., a set for which $M(\mathcal{A}) = \mathcal{A}$) that attracts each bounded set of R_+^m and is maximal (i.e., every compact invariant set of M belongs to $\mathcal{A}$).

Since $\delta < 1$, there exists an integer $t' \geq 1$ (depending on $x(0)$) such that

$$|x(t)| \leq 2\frac{f}{1-\delta}, \quad t \in I[t', +\infty),$$

which implies (1.23) with $c \doteq 2f(1-\delta)^{-1}$. □

The condition (1.24) implies that at each time step, and for any class distribution x, there is loss of individuals during the transition from any class to any other class (due to, say, mortality). Condition (1.25) implies that there is an upper bound to fertility. For every $i, j \in I[1, m]$ it requires $f_{ij}(x)x_j$ to be bounded for $x \in R_+^m$, i.e., that there be intraclass density effects on fertility.

By Theorem 1.2.1 the extinction equilibrium $x = 0$ of (1.20) loses stability as r increases through the critical value 1. Recall for linear matrix equations that there exists an unbounded continuum of nonextinction equilibria that "bifurcates" at the point $(r, x) = (1, 0)$, but that the spectrum for these nonextinction equilibria consists of the single point $r = 1$ (Fig. 1.1). The bifurcation of nonextinction equilibria from this same point for the nonlinear matrix equation (1.20) will be studied in section 1.2.3.

From Theorem 1.1.3 it follows that r can be replaced by the inherent ne reproductive number n in the statement of Theorem 1.2.1, provided $(I - T)^{-1}u > 0$.

Theorem 1.2.1 can be applied to all of the examples (1.16), (1.17), (1.19), and (1.18) above. In each case, $r < 1$ (or $n < 1$) implies global extinction and $r > 1$ ($n > 1$) implies uniform persistence with respect to $x = 0$.

1.2.2 Matrix equations with parameters. Theorem 1.2.1 implies that if a model parameter λ appearing in the projection matrix of a nonlinear matrix equation (1.20) is changed in such a way as to cause the dominant eigenvalue r of the inherent projection matrix to increase through the critical value 1, then the extinction equilibrium $x = 0$ will lose its (local asymptotic) stability. This suggests a possible bifurcation of nontrivial equilibria $x \neq 0$ from the extinction equilibrium $x = 0$ at the critical value λ_0 of λ, where $r = 1$. For this scenario to happen, the model parameter λ must be present (or related to entries) in the inherent projection matrix $P(0)$. Call such a parameter an *inherent parameter*. In applications there are usually a considerable number of inherent model parameters from which to choose. The choice can be dictated by biological considerations related to the biological problem of interest or by mathematical considerations as to which parameter is important with regard to describing the dynamics of the specific model under consideration. One parameter that can be used is the inherent net reproductive number n. While this number generally does not appear explicitly in the projection matrix, it can be introduced by scaling the class specific fertilities to n by writing $f_{ij} = n\varphi_{ij}$. The projection matrix can be written

$$P(n, x) = T(x) + n\Phi(x), \tag{1.26}$$

where $\Phi(x) \geq 0$ is the *normalized fertility matrix.* Φ is normalized in the sense that 1 is the dominant eigenvalue of $T(0) + \Phi(0)$ and hence the dominant eigenvalue of $\Phi(0)\,(I - T(0))^{-1}$. (Note that $0 \neq \Phi(0) \geq 0$.) With this prototype in mind we will consider matrix equations in which the parameter λ appears linearly in the inherent projection matrix.

Consider the matrix equation

$$
\begin{aligned}
&x(t+1) = P(\lambda, x(t))\, x(t), \\
&P(\lambda, x) = [p_{ij}(\lambda, x)], \quad p_{ij} \in C^{k+1}\left(R^1 \times \Omega, R^1_+\right), \quad k \geq 0.
\end{aligned}
\tag{1.27}
$$

Assume that

$$
P(\lambda, 0) = A + \lambda B \tag{1.28}
$$

for some $m \times m$ matrices A and B.

DEFINITION 3. *A real number $\lambda_0 \in R^1$ is called a* critical value *of $A + \lambda B$ if 1 is an algebraically simple eigenvalue of $A + \lambda_0 B$ and there exist left and right eigenvectors $w \geq 0$ and $v > 0$, respectively, such that $w^\tau Bv \neq 0$. If, in addition, 1 is a strictly dominant eigenvalue of $A + \lambda_0 B$, then λ_0 is called a* strictly dominant critical value.

Note that since w is an eigenvector, $w \neq 0$ and it follows that $w^\tau v > 0$. Suppose that λ_0 is a critical value of $P(\lambda, 0) = A + \lambda B$. If $c(r, \lambda)$ denotes the characteristic polynomial $\det(rI - A - \lambda B)$ of $P(\lambda, 0)$, then $c(1, \lambda_0) = 0$ and $\partial c(1, \lambda_0)/\partial r \neq 0$ (since 1 is simple). The implicit function theorem implies that there is a simple (infinitely differentiable) root $r = r(\lambda)$ of $c(r, \lambda)$ such that $r(\lambda_0) = 1$. Let $v = v(\lambda)$ denote the associated (infinitely differentiable) right eigenvector. A differentiation of

$$
(A + \lambda B)\, v(\lambda) = r(\lambda) v(\lambda)
$$

leads, after evaluation at $\lambda = \lambda_0$, to

$$
(A + \lambda_0 B)\, v'(\lambda_0) = v'(\lambda_0) + (r'(\lambda_0) v - Bv)\,.
$$

Since $I - (A + \lambda_0 B)$ is singular, it follows that $w^\tau\,(r'(\lambda_0)v - Bv) = 0$, and consequently

$$
r'(\lambda_0) = \frac{w^\tau B v}{w^\tau v}.
$$

We see that $r(\lambda)$ is increasing or decreasing at λ_0 according to the sign of $w^\tau Bv$.

Using linearized stability theory at $x = 0$ (see Appendix A.2) together with the fact that the Jacobian of $P(\lambda, x)x$ at $x = 0$ is equal to the inherent projection matrix $P(\lambda, 0)$, we obtain the following result.

THEOREM 1.2.3. *Assume that λ_0 is a strictly dominant critical value of $A + \lambda B$. If $w^\tau Bv > 0$, then the extinction equilibrium $x = 0$ of (1.27)–(1.28) loses stability as λ is increased through λ_0. If $w^\tau Bv < 0$, then the extinction equilibrium $x = 0$ loses stability as λ is decreased through λ_0.*

For the case (1.26) $w^\tau Bv = w^\tau \Phi(0) v > 0$. Theorems 1.1.3 and 1.2.3 imply that $x = 0$ loses stability as n is increased through 1.

1.2.3 Positive equilibria. The equilibrium equation associated with (1.27)–(1.28) is

$$x = P(\lambda, x)x. \tag{1.29}$$

A solution pair (λ, x) of this equation will be called an *equilibrium pair.* An equilibrium pair $(\lambda, 0)$ is called an *extinction equilibrium pair.* An equilibrium pair (λ, x) is *positive* if $x > 0$; it is *nonnegative* if $x \geq 0$. A *nonextinction* equilibrium pair is an equilibrium pair (λ, x) with $x \in R^m_+/\{0\}$. A *boundary* equilibrium pair is an equilibrium pair (λ, x) with $x \in \partial R^m_+$. Thus, for a nonextinction boundary equilibrium pair (λ, x) at least one component of $x \in R^m_+/\{0\}$ is equal to 0. A *continuum* of equilibrium pairs is a closed and connected set in $R^1 \times R^m$.

We can write the projection matrix in equation (1.27) as

$$P(\lambda, x) = P(\lambda, 0) + R(\lambda, x),$$

where the entries in the matrix $R(\lambda, x) \doteq P(\lambda, x) - P(\lambda, 0)$ are continuous and $O(|x|)$ near $x = 0$. The equilibrium equation becomes

$$x = (A + \lambda B)\, x + r(\lambda, x),$$

where $r(\lambda, x) \doteq R(\lambda, x)x$ is continuous and $|r(\lambda, x)| = O(|x|^2)$ near $x = 0$. If $I - A$ is invertible, then the equilibrium equation can be written equivalently as

$$x = \lambda L x + h(\lambda, x),$$

$$L \doteq (I - A)^{-1}B, \quad h(\lambda, x) \doteq (I - A)^{-1}r(\lambda, x),$$

where $h(\lambda, x)$ is continuous and $|h(\lambda, x)| = O(|x|^2)$ near $x = 0$. This equation has the form (B.2) in Appendix B to which the theorems of Appendix B can be applied. A bifurcation of nontrivial equilibria can only occur at characteristic values of L. According to Theorem B.1.1 (see Appendix B) a continuum of positive equilibria bifurcates from the extinction equilibrium at any (geometrically) simple characteristic value of L associated with a positive characteristic vector.

LEMMA 1.2.1. *Suppose that λ_0 is a critical value of $A + \lambda B$ and $w \geq 0$ and $v > 0$ are left and right eigenvectors of $A + \lambda_0 B$. If $I - A$ is invertible, then*

(a) *λ_0 is a geometrically simple characteristic value of $L = (I - A)^{-1}B$ with right characteristic vector $v > 0$;*

(b) *a left characteristic vector q of L is given by $q^\tau = w^\tau(I - A)$;*

(c) *$q^\tau v = w^\tau(I - A)v = \lambda_0 w^\tau Bv$.*

Proof. The equation $(A + \lambda_0 B)v = v$ is equivalent to $\lambda_0 L v = v$. Since the null space of $I - (A + \lambda_0 B)$ is spanned by v, the same is true of the null space of $I - \lambda_0 L$. This proves (a). From the definition of a critical value λ_0 we have $w^\tau(A + \lambda_0 B) = w^\tau$, and hence $\lambda_0 w^\tau B = w^\tau(I - A) = q^\tau$. From this follows (c). To show (b) we note that $q^\tau \lambda_0 L = w^\tau(I - A)\lambda_0(I - A)^{-1}B = \lambda_0 w^\tau B = q^\tau$. □

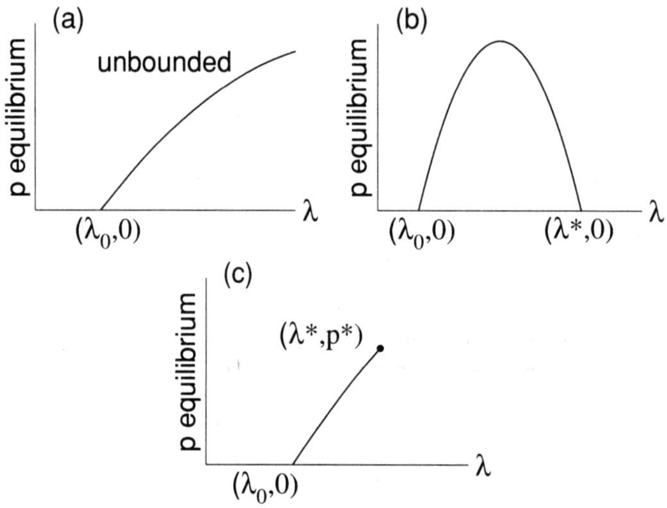

FIG. 1.2. *The three alternatives in Theorem 1.2.4 for the global bifurcating continuum C^+ of positive equilibria are schematically represented by graphs of equilibrium total population size p against λ. In* (c) *p^* is the total population size of a nonextinction boundary equilibrium x^*.*

The following theorem follows immediately from this lemma and Theorem B.1.1. It states, roughly, that as λ is varied a "branch" of positive equilibria bifurcates from $x = 0$ at a critical value λ_0 and connects to the boundary of the nonnegative cone R^m_+ (∞ is included in the boundary).

THEOREM 1.2.4. *Suppose that $k \geq 1$ in* (1.27)–(1.28), *and suppose that $I - A$ is invertible. Suppose that λ_0 is a critical value of $A + \lambda B$. Then there exists a continuum C^+ of equilibrium pairs such that $(\lambda_0, 0) \in C^+$ and $C^+/\left\{R^1 \times \partial R^m_+\right\} \neq \emptyset$ contains only positive equilibria. Furthermore, one of the following alternatives holds:*

(a) *$C^+/\left\{(\lambda_0, 0)\right\}$ is unbounded in $R^1 \times R^m_+$ and contains only positive equilibrium pairs;*

(b) *C^+ contains a point $(\lambda^*, 0)$, where $\lambda^* \neq \lambda_0$ is a characteristic value of $(I - A)^{-1}B$ with a nonnegative characteristic vector;*

(c) *C^+ contains a nonextinction boundary equilibrium $(\lambda^*, x^*) \in R^1 \times \partial R^m_+$.*

The three alternatives in Theorem 1.2.4 are illustrated graphically in Fig. 1.2. In applications it is often the case that alternatives (b) and (c) can be ruled out, with the result that the bifurcating branch is positive and unbounded as in (a). For example, a case that often occurs is that L has only one characteristic value λ_0 (or at least no other with a nonnegative characteristic vector). In this case, alternative (b) is ruled out.

It also frequently occurs that the only nonnegative solution of the equilibrium equation (1.29) is $x = 0$. In this case, alternative (c) is ruled out.

For example, a nonextinction equilibrium $x \geq 0$ is an eigenvector of the projection matrix $P(\lambda, x)$ associated with eigenvalue 1. If the matrix $P(\lambda, x)$ is irreducible and primitive for all $(\lambda, x) \in R^1 \times R_+^m$, then it can have no nonnegative eigenvectors other than the positive eigenvector associated with the dominant eigenvalue. It follows that $x > 0$ (and 1 is the dominant eigenvalue). Thus, for such projection matrices alternative (c) is ruled out. For an application in which alternative (c) does occur, however, see the spotted owl model in section 1.3.

Analytic formulas for positive equilibria of nonlinear matrix equations, expressed in terms of the model's parameters, are not in general available. However, general approximations can be obtained for the positive equilibrium pairs on the continuum C^+ near the bifurcation point $(\lambda_0, 0)$. This can be done by the classical procedure of parameterizing the branch of equilibrium pairs $(\lambda, x) = (\lambda(\varepsilon), x(\varepsilon))$, $(\lambda(0), x(0)) = (\lambda_0, 0)$, in terms of a small parameter $\varepsilon \approx 0$ (sometimes referred to as a Liapunov/Schmidt expansion). The details of this procedure and a proof of its validity are given in Theorem B.2.1 (see Appendix B.2). An application of that theorem to the matrix equation (1.27)–(1.28) results in the theorem below. Here use is made of Lemma 1.2.1. Define $V^\perp = \{w \in R^m \mid w^\tau v = 0\}$, and let G be the matrix such that $x = Gf$ is the unique solution of the linear algebraic equation $x = \lambda_0 (I - A)^{-1} Bx + f$ lying in $V^\perp$.

THEOREM 1.2.5. *Suppose that $k \geq 1$ in (1.27)–(1.28), and suppose that $I - A$ is invertible. Suppose that λ_0 is a critical value of $A + \lambda B$ and $w^\tau Bv \neq 0$. Then in a sufficiently small neighborhood of the bifurcation point $(\lambda_0, 0)$ the continuum C^+ has the form*

$$\lambda = \lambda_0 + \eta(\varepsilon), \quad x = \varepsilon v + z(\varepsilon), \quad \varepsilon > 0, \tag{1.30}$$

where, for some $\varepsilon_0 > 0$,

$$\begin{aligned} &\eta \in C^k\left((-\varepsilon_0, \varepsilon_0), R^1\right), \quad z \in C^k\left((-\varepsilon_0, \varepsilon_0), V^\perp\right), \\ &\eta(0) = 0, \quad z(0) = z'(0) = 0. \end{aligned}$$

If $k \geq 2$, then

$$\begin{aligned} &\lambda = \lambda_0 + \lambda_1 \varepsilon + O(\varepsilon^2), \quad x = \varepsilon v + z_1 \varepsilon^2 + O(\varepsilon^3), \\ &\lambda_1 = -\frac{w^\tau [d_{ij}] v}{w^\tau Bv}, \quad z_1 = G\left([d_{ij}] v + \lambda_1 v\right), \end{aligned} \tag{1.31}$$

where $d_{ij} = \left(\nabla_x p_{ij}(\lambda_0, 0)\right)^\tau v$.

The formula for λ_1 in (1.31) allows for a determination of the *direction of bifurcation* of C^+, i.e., the sign of $\lambda - \lambda_0$ near the bifurcation point. If $\lambda > \lambda_0$ $(\lambda < \lambda_0)$ for $(\lambda, x) \in C^+$ near $(\lambda_0, 0)$, we say that the bifurcation is *to the right (to the left)*. The direction of bifurcation is determined by the sign of λ_1; if $\lambda_1 > 0$ $(\lambda_1 < 0)$, then the bifurcation is to the right (to the left). The coefficient z_1 describes the effect the nonlinearities have on the equilibrium class distribution x as a perturbation from the inherent (linearized) distribution v.

The local stability of an equilibrium x of equation (1.27) is determined by the eigenvalues of the Jacobian $J(\lambda, x)$ of the right-hand side $P(\lambda, x)x$ evaluated at x. Under the conditions of Theorem 1.2.5, the positive equilibrium pairs near the bifurcation point $(\lambda_0, 0)$ are parameterized by ε as in (1.31). For these equilibria the Jacobian $J(\lambda, x) = A + \lambda B + Jr(\lambda, x)$, $Jr(\lambda, x) = [\partial_j r_i(\lambda, x)]$, and hence its eigenvalues, are functions of the parameter ε. A proof of the following lemma appears in Appendix C.

LEMMA 1.2.2. *Suppose that* $k \geq 2$ *in* (1.27)–(1.28), *and suppose that* $I - A$ *is invertible. Suppose that* λ_0 *is a critical value of* $A + \lambda B$ *and* $w^\tau Bv \neq 0$. *Let* $\zeta(\varepsilon)$ *be the eigenvalue of the Jacobian* $J(\lambda, x) = A + \lambda B + Jr(\lambda, x)$ *evaluated at the nontrivial equilibria* (1.31) *such that* $\zeta(0) = 1$. *Then*

$$\zeta'(0) = -\frac{w^\tau Bv}{w^\tau v}\lambda_1,$$

where λ_1 *is given in* (1.31).

If 1 is a strictly dominant eigenvalue of the Jacobian at $(\lambda_0, 0)$, then the stability of the positive equilibria (1.31) from the bifurcating continuum C^+ in Theorem 1.2.5 near the bifurcation point $(\lambda_0, 0)$ is determined by the eigenvalue $\zeta(\varepsilon)$ for small ε; i.e., the positive equilibria are (locally asymptotically) stable if $\zeta(\varepsilon) < 1$ and unstable if $\zeta(\varepsilon) > 1$ for small $\varepsilon > 0$. Thus, local stability is determined by the sign of $\zeta'(0)$. By Lemma 1.2.2 this sign is determined by the sign of λ_1 and hence (by Theorem 1.2.5) by the direction of bifurcation.

The local bifurcation described in Theorem 1.2.5 will be called a *stable bifurcation* if the positive equilibria (1.30) are (locally asymptotically) stable in a sufficiently small neighborhood of the bifurcation point $(\lambda_0, 0)$ (i.e., for $\varepsilon > 0$ sufficiently small). If these positive equilibria are unstable, then the bifurcation will be called *unstable.* The next theorem follows immediately from Lemma 1.2.2.

THEOREM 1.2.6. *Suppose that* $k \geq 2$ *in* (1.27)–(1.28), *and suppose that* $I - A$ *is invertible. Suppose that* λ_0 *is a strictly dominant critical value of* $A + \lambda B$. *Assume that* $\lambda_1 \neq 0$, *where* λ_1 *is given by* (1.31).

If $w^\tau Bv > 0$, *then the bifurcation described in Theorem* 1.2.5 *is stable if it is to the right and unstable if it is to the left.*

If $w^\tau Bv < 0$, *then the bifurcation described in Theorem* 1.2.5 *is stable if it is to the left and unstable if it is to the right.*

As a result of this theorem the stability or instability of the bifurcation at $(\lambda_0, 0)$ can be determined from the direction of bifurcation alone and an equilibrium stability analysis is unnecessary. The four possibilities are illustrated in Fig. 1.3.

1.2.4 Positive equilibrium destabilization. The stability property of the bifurcating continuum C^+ of equilibria in Theorem 1.2.6 may not persist outside of a neighborhood of the bifurcation point $(\lambda_0, 0)$. The stable positive equilibria arising from a stable bifurcation at $(\lambda_0, 0)$ may lose their stability as one "moves away" from $(\lambda_0, 0)$ along C^+. Or, conversely, the unstable positive equilibria arising from an unstable bifurcation may gain stability.

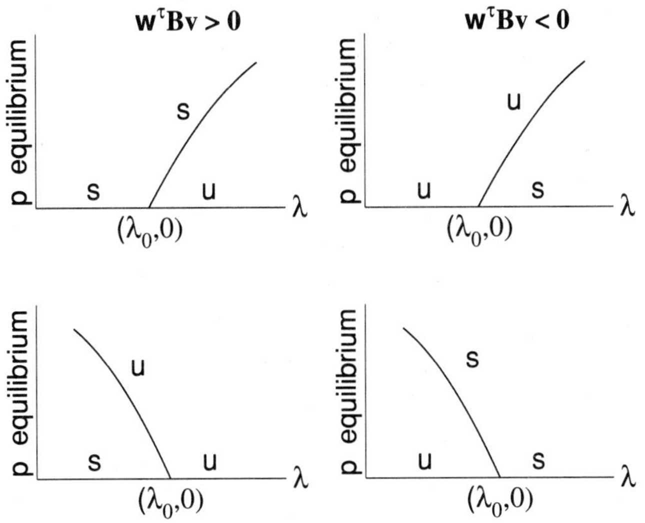

FIG. 1.3. *The local bifurcation alternatives for the branch of positive equilibria in Theorem 1.2.6 are schematically represented by graphs of total population size p against the parameter* λ. *The letter "s" denotes "stable" and "u" denotes "unstable."*

Equilibrium destabilization will occur as one moves along C^+ if the magnitude of an eigenvalue of the Jacobian becomes greater than 1; i.e., an eigenvalue moves from the inside to the outside of the unit circle in the complex number plane. When this happens, another attractor usually appears and a "bifurcation" is said to occur. What kind of new attractor arises depends crucially on the point at which the eigenvalue leaves the unit complex circle. General theorems describing the bifurcation possibilities for general maps appear in most textbooks on discrete dynamical systems (e.g., see [193], [436]). Nothing exceptional can be said about matrix population models (1.27) as a specialized class of higher order maps. We will therefore only describe, for future reference, the most common general local bifurcations that occur for higher order maps.

Stability is lost at an equilibrium x_{cr} as the parameter λ is increased (or decreased) through a critical value λ_{cr} because an eigenvalue ζ of the Jacobian at x_{cr} moves out of the complex unit circle. That is to say, $|\zeta| = 1$ for $\lambda = \lambda_{cr}$ while, in a neighborhood of λ_{cr}, $|\zeta| < 1$ for $\lambda < \lambda_{cr}$ and $|\zeta| > 1$ for $\lambda > \lambda_{cr}$ (or vice versa, $|\zeta| < 1$ for $\lambda > \lambda_{cr}$ and $|\zeta| > 1$ for $\lambda < \lambda_{cr}$). Moreover, it is usually the case that no other eigenvalue leaves the unit circle at $\lambda = \lambda_{cr}$ if ζ is real; or only ζ and its complex conjugate leave at $\lambda = \lambda_{cr}$ if ζ is complex. If ζ is real, then either $\zeta = 1$ or -1 at $\lambda = \lambda_{cr}$.

When $\zeta = 1$ at $(\lambda, x) = (\lambda_{cr}, x_{cr})$, the bifurcation typically involves equilibria only. There are a variety of possibilities, the most fundamental of which are the *transcritical*, *saddle-node*, and *pitchfork* bifurcations. A transcritical bifurcation is the crossing of two distinct branches of equilibria at the bifurcation

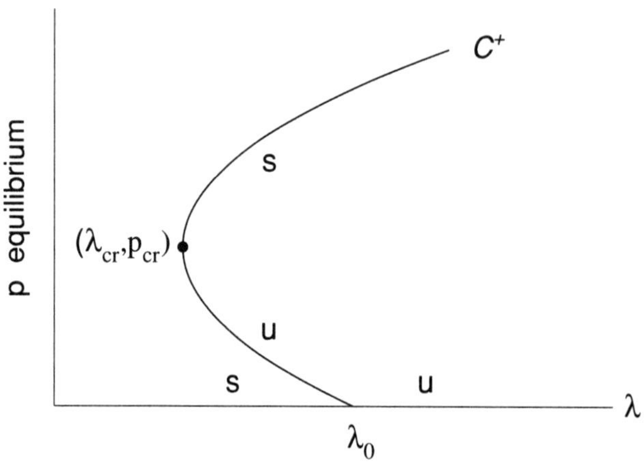

FIG. 1.4. *A transcritical bifurcation occurs at the point* $(\lambda, p) = (\lambda_0, 0)$ *where the extinction equilibrium* $p = 0$ *loses stability. A saddle-node bifurcation occurs at the point* $(\lambda, p) = (\lambda_{cr}, p_{cr})$. *The letter "s" denotes "stable" and "u" denotes "unstable."*

point in such a way that equilibria from both branches exist for λ values on both sides of λ_{cr}. A transcritical bifurcation usually involves, at the bifurcation point, an *exchange of stability* from the equilibria on one branch to those on the other. The bifurcation of positive equilibria from the extinction equilibrium $(\lambda_{cr}, x_{cr}) = (\lambda_0, 0)$ as described in Theorem 1.2.5 is a transcritical bifurcation (the other "half" of the branch corresponding to $\varepsilon < 0$ is ignored because it consists of negative equilibrium pairs and hence is not biologically relevant). Transcritical bifurcations at positive equilibria $x_{cr} > 0$ do not typically occur in population models.

Saddle-node bifurcations at positive equilibria, on the other hand, do occur in population models. A saddle-node bifurcation is one in which, at least in a neighborhood of the bifurcation point (λ_{cr}, x_{cr}), two positive equilibria exist for λ on one side of λ_{cr} while no positive equilibria exist on the other side of λ_{cr}. Thus, at a saddle-node bifurcation point (λ_{cr}, x_{cr}) the continuum C^+ "turns around." Usually (but not always) the two positive equilibria have opposite stability properties, one being a saddle and the other a stable node. One circumstance in which a saddle-node bifurcation occurs in population models is when the bifurcation at $(\lambda, x) = (\lambda_0, 0)$ is unstable (due, for example, to an Allee effect), but the continuum C^+ "turns around" at a saddle-node bifurcation to create stable equilibria. See Fig. 1.4. The spotted owl model in section 1.3 is a specific example.

Finally, in a neighborhood of pitchfork bifurcation point (λ_{cr}, x_{cr}), three equilibria occur on one side of λ_{cr} while only one equilibrium occurs on the other side. Pitchfork bifurcations have rarely occurred in population models.

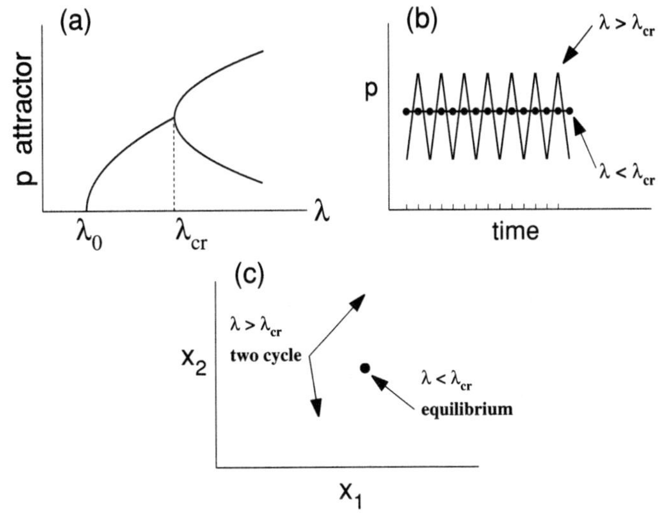

FIG. 1.5. *As λ is increased through λ_{cr} the positive equilibrium is destabilized and a stable 2-cycle emerges as shown in the bifurcation diagram and time series plots in graphs* (a) *and* (b). *In phase space the attractor changes from an equilibrium point to a pair of points on the 2-cycle* (*shown in graph* (c) *for $m = 2$ dimensions*).

The second possibility $\zeta = -1$ at $(\lambda, x) = (\lambda_{cr}, x_{cr})$ usually implies the bifurcation of a branch of 2-cycles, i.e., periodic solutions of period 2. This "period doubling" bifurcation, familiar for one-dimensional maps, also occurs for higher dimensional maps. These 2-cycles "grow" or "pop" out of the equilibrium at the bifurcation point, starting with a small amplitude that increases as λ moves further from λ_{cr}. The 2-cycles exist for λ on one side of λ_{cr}, but not the other, and at such a value of λ the 2-cycle has the opposite stability of the equilibrium. If the 2-cycles are stable (i.e., exist on the same side of λ_0 as do the unstable equilibria), the bifurcation is called *stable or supercritical.* In the opposite case when the 2-cycles are unstable (i.e., exist on the same side of λ_0 as do the stable equilibria), the bifurcation is *unstable or subcritical.* See Fig. 1.5 for an illustration of a stable 2-cycle bifurcation.

The final possibility $\zeta = \exp(\pm i\theta)$, $\theta \neq 0$ or π, at $(\lambda, x) = (\lambda_{cr}, x_{cr})$ can only occur for maps of dimension $m = 2$ or higher and therefore are particularly relevant for structured population models. In this case a so-called *invariant loop* (usually) bifurcates from the equilibrium. This loop is a one-dimensional, closed curve in R^m (roughly elliptical near the bifurcation point). It is not itself a solution or orbit, but it is an invariant set of points. The orbits lying on the invariant loop may either be "quasi periodic" or "period locked"; that is to say, they may either move around the loop without being mathematically periodic (and hence never hitting a point twice), or they may be periodic in their motion around the loop. Invariant loops are best seen in phase space R^m rather than

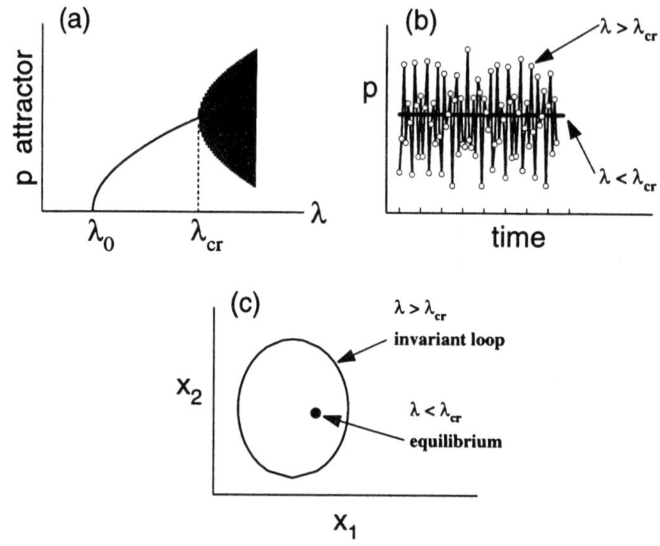

FIG. 1.6. *As λ is increased through λ_{cr} the positive equilibrium is destabilized and quasi-periodic oscillations emerge as shown in the bifurcation diagram and time series plots in graphs* (a) *and* (b). *In phase space the attractor changes from an equilibrium point to a closed invariant loop (shown in graph* (c) *for $m = 2$ dimensions).*

in time series plots of components of x against time t. See Fig. 1.6. The invariant loop bifurcation, sometimes called a *discrete Hopf (or Naimark/Sacker) bifurcation*, requires the unexpected technical assumption that ζ not be equal to any of the first four roots of unity; at such points the nature of the bifurcation is apparently not yet fully understood. The invariant loop bifurcation may be either stable or unstable; in other words, orbits starting near the loop may tend towards or away from the loop in forward time.

The analytical determination of the type and stability properties of a bifurcation is usually difficult and intractable in applications, although it is in principle possible by means of known procedures [306].

1.2.5 The net reproductive number. In this section we take a closer look at using the inherent net reproductive number n as the bifurcation parameter λ.

Consider the general projection matrix $P(x) = T(x) + F(x) \geq 0$. Assume that $T(x)+F(x)$ is irreducible and primitive and $I-T(x)$ is nonsingular for all $x \in R_+^m$. If $F(x)(I-T(x))^{-1}$ has a positive, strictly dominant, simple eigenvalue $n = n(x)$ with a nonnegative eigenvector $v = v(x) \geq 0$, then $n(x)$ is called the *net reproductive number at* x. The quantity $n(x)$ is not to be confused with the *inherent* net reproductive number n (which is, in fact, $n = n(0)$). Suppose that $x > 0$ is a positive equilibrium so that x solves the equation $x = (T(x)+F(x))x$. Since $x > 0$ and the matrix $T(x) + F(x)$ is irreducible and primitive, it follows that 1 is the dominant eigenvalue of $T(x) + F(x)$. From Theorem 1.1.3 we conclude

$$n(x) = 1 \quad \text{for all equilibria} \quad x > 0. \tag{1.32}$$

The biological interpretation of this statement is, roughly, that at equilibrium each individual exactly replaces itself over the course of its lifetime.

If the fertilities f_{ij} are scaled to the inherent net reproductive number n, $f_{ij} = n\varphi_{ij}$, then we can write $F(x) = n\Phi(x)$, where $\Phi(x) \geq 0$ $(\neq 0)$ is the normalized fertility matrix, and the matrix equation (1.27) becomes

$$x(t+1) = (T(x(t)) + n\Phi(x(t)))\, x(t). \tag{1.33}$$

We consider this equation under the following assumptions. For some integer $k \in I[0, +\infty)$

(1.34)

(a) $T(x) = [t_{ij}(x)]$, $t_{ij} \in C^{k+1}(\Omega, [0,1])$, and $\sum_{i=1}^{m} t_{ij}(x) \leq 1$ for each $j \in I[1, m]$;

(b) $I - T(x)$ is invertible for all $x \in R_+^m$;

(c) $\Phi(x) = [\varphi_{ij}(x)]$, $\varphi_{ij} \in C^{k+1}(\Omega, R_+^1)$;

(d) $T(x) + n\Phi(x)$ is irreducible and primitive for all $x \in R_+^m$, $n > 0$;

(e) for all $x \in R_+^m$, $\Phi(x)(I - T(x))^{-1}$ has a positive, strictly dominant, simple eigenvalue value $\nu(x)$, $\nu(0) = 1$, with a nonnegative eigenvector vector $u(x) \geq 0$, such that $(I - T(x))^{-1}u(x) > 0$.

The last assumption implies that $n(x) = n\nu(x)$, where n is the inherent net reproductive number. From (1.32)

$$n\nu(x) = 1 \quad \text{for all equilibria} \quad x > 0. \tag{1.35}$$

By Theorems 1.1.3 and 1.2.1, $x = 0$ loses stability as n is increased through the critical value 1. From Theorems 1.2.5 and 1.2.6 the bifurcation of positive equilibria at $n = 1$ is stable if the bifurcation is to the right and unstable if it is to the left. Define the *spectrum* and *range* of the continuum C^+ from Theorem 1.2.4 as

$$\begin{aligned} \sigma(C^+) &\doteq \{n \mid (n, x) \in C^+\} \subset R^1, \\ \rho(C^+) &\doteq \{x \mid (n, x) \in C^+\} \subset R_+^m, \end{aligned}$$

respectively. The spectrum and the range are connected sets. For linear models $s(C^+) = \{1\}$ and $\rho(C^+) = \{cv \mid c \in R_+^m\}$.

THEOREM 1.2.7. *Consider the matrix equation* (1.33) *under the conditions* (1.34). *Then*

(i) $\rho(C^+)$ *is unbounded and* $\rho(C^+)/\{0\}$ *contains only positive equilibria;*

(ii) $\sigma(C^+)$ *contains only positive* $n > 0$.

Proof. (i) First we rule out alternatives (b) and (c) in Theorem 1.2.4.

Suppose that $(n_1, 0) \in C^+$, where $n_1 \neq 1$ is a characteristic value of $L = (I - T(0))^{-1}\Phi(0)$ (and hence of $\Phi(0)(I - T(0))^{-1}$) with a nonnegative characteristic vector $\xi \geq 0$. Then 1 is an eigenvalue and ξ is an associated eigenvector of the matrix $T(0) + n_1\Phi(0)$. Since $T(0) + n_1\Phi(0)$ is irreducible and primitive, a nonnegative eigenvector can be associated only with the dominant eigenvalue. Therefore 1 is the dominant eigenvalue of this matrix. By Theorem 1.1.3 it follows that the dominant eigenvalue of $n_1\Phi(0)(I - T(0))^{-1}$ is equal to 1, which by (1.34(e)) implies the contradiction $n_1 = 1$. This rules out alternative (b).

By (1.34(d)) the projection matrix $T(x) + n\Phi(x)$ can have no nonnegative eigenvector other than the positive eigenvector associated with the dominant eigenvalue. Since an equilibrium is an eigenvector of the projection matrix associated with eigenvalue 1, it follows that a nonnegative equilibrium must in fact be positive. This rules out alternative (c).

By alternative (a) of Theorem 1.2.4 $\rho(C^+)$ contains only positive equilibria and either it or the spectrum is unbounded. If the range $\rho(C^+)$ were bounded, then $\nu(x) > 0$ would be bounded away from 0 for all $x \in \rho(C^+)$ and (1.35) would imply that the spectrum $\sigma(C^+)$ would also be bounded, a contradiction. This proves (i).

(ii) Suppose that there exists a pair $(n, x) \in C^+$ for which $n \leq 0$. Because C^+ is a continuum, it follows that there must exist an equilibrium pair $(0, x) \in C^+$. By (i), $x > 0$. From the equilibrium equation it follows that $x = T(x)x$ which contradicts (1.34(b)). Thus, $n > 0$ for all $(n, x) \in C^+$. □

The next theorem contains facts about the spectrum obtainable from properties of $\nu(x)$ (or equivalently $n(x)$).

THEOREM 1.2.8. *Consider the matrix equation* (1.33) *under the conditions* (1.34).

(i) *If* $\lim_{|x| \to +\infty,\ x \in R^m_+} \nu(x) = 0$, *then there exists at least one positive equilibrium for each* $n > 1$.

(ii) *If* $\nu(x) < 1$ *for* $x > 0$, $x \approx 0$, *then the bifurcation is to the right and stable. If* $\nu(x) > 1$ *for* $x > 0$, $x \approx 0$, *then the bifurcation is to the left and unstable.*

(iii) *If* $\nu(x) < 1$ *for all* $x > 0$, *then there exists no positive equilibrium for* $n < 1$.

Proof. If $\lim_{|x| \to +\infty,\ x \in R^m_+} \nu(x) = 0$, the unboundedness of the range $\rho(C^+)$ and (1.35) imply that the spectrum $\sigma(C^+)$ is unbounded; i.e., $\sigma(C^+)$ is an unbounded interval of positive real numbers that contains 1 in its closure. This implies (i). (Note that it is sufficient that $\nu(x)$ tends to 0 for $x \in \rho(C^+)$.) Both (ii) and (iii) follow immediately from (1.35). □

In applications $\nu(x) \to 0$ as $|x| \to +\infty$ is usually a consequence of the modeling assumption that fertility and/or transition (e.g., survival) rates tend to 0 as $|x|$ increases without bound.

In applications, the inequality $\nu(x) < 1$, $x > 0$, is usually the result of the modeling assumption that all density effects are deleterious, i.e., that fertility and/or transition rates are, in the presence of other individuals, less than inherent values. This condition will be fulfilled, for example, if for all $0 < x \in R_+^m$

$$0 \le \Phi(x) < \Phi(0), \quad 0 \le T(x) < T(0). \tag{1.36}$$

To see this, note that from $0 \le T(x)y < T(0)y$, $0 \ne y \in R_+^m$, we obtain $0 \le T^i(x)y < T^i(0)y$ and hence

$$0 \le (I - T(x))^{-1} y = \left(I + T(x) + T^2(x) + \cdots\right) y < (I - T(0))^{-1} y.$$

Then $0 \le \Phi(x)(I - T(x))^{-1} y < \Phi(0)(I - T(0))^{-1} y$, and since the unit sphere is compact in R^m,

$$\begin{aligned} \nu(x) &= \max_{|y|=1,\ y \in R_+^m} \left|\Phi(x)(I - T(x))^{-1} y\right| \\ &< \max_{|y|=1,\ y \in R_+^m} \left|\Phi(0)(I - T(0))^{-1} y\right| = \nu(0). \end{aligned}$$

Often in applications density effects are modeled by a dependence of the fertility matrix $F = F(p)$ and/or the transition matrix $T = T(p)$ on a *weighted total population size*

$$p = \sum_{i=1}^m \omega_i x_i, \quad \omega_i \ge 0, \quad \sum_{i=1}^m \omega_i \ne 0.$$

Then $\nu = \nu(p)$ is a scalar function of a scalar variable and (1.35) describes a curve in the (n, p) plane. A plot of this planar curve is one way of geometrically describing the bifurcation of the continuum C^+. For example, from (1.35) we see that $\nu'(0) < 0$ implies a right and stable bifurcation and $\nu'(0) > 0$ implies a left and stable bifurcation.

THEOREM 1.2.9. *In* (1.33)–(1.34) *suppose that* $F = F(p)$ *and* $T = T(p)$ *are functions of a weighted total population size* $p = \sum_{i=1}^m \omega_i x_i$, $\omega_i \ge 0$, $\sum_{i=1}^m \omega_i \ne 0$. *If* $\nu'(p) < 0$ *for all* $p > 0$, *then there exists no positive equilibrium for* $n < 1$ *and* $\sigma(C^+) = [1, 1/\nu_\infty)$, *where* $\nu_\infty = \lim_{p \to +\infty} \nu(p) \ge 0$ *(replace* $1/\nu_\infty$ *by* $+\infty$ *if* $\nu_\infty = 0$*). Moreover, for each* $n \in \sigma(C^+)$ *there exists exactly one positive equilibrium* x.

Proof. All that needs to be proved is the uniqueness statement; the rest of the theorem follows from (1.35). Suppose that $x > 0$ and $x^* > 0$ are equilibria associated with the same value of n. By (1.35) x and x^* have the same weighted total population size p and hence both are positive eigenvectors, associated with eigenvalue 1, of the same matrix $T(p) + F(p) = T(p) + n\Phi(p)$. By assumption (1.34(d)) this matrix is irreducible and primitive and therefore has a only one independent positive eigenvector. Thus, x and x^* are dependent. In fact, they must be identical since they have the same weighted total population size. $\square$

A similar argument shows that if $\nu'(p) > 0$ for all $p > 0$, then there exists no positive equilibrium for $n > 1$ and $\sigma(C^+) = [1/v_\infty, 1)$. Moreover, for each $n < 1$, $n \in \sigma(C^+)$, there exists exactly one positive equilibrium x.

A positive derivative $\nu'(p) > 0$ can arise from an "Allee" effect in which some fertility and/or transition rates increase with increased total population size. Allee effects are usually postulated to hold at low population densities, while the usual deleterious density effects are assumed to hold at high population densities. In such a model, $\nu'(p) > 0$ for small p, say $0 < p < p_{cr}$, and $\nu'(p) < 0$ for large $p > p_{cr}$. Using (1.35) we can deduce a bifurcation diagram as in Fig. 1.4. Thus, in this case there exist two positive equilibria for n satisfying $n_{cr} = 1/\nu(p_{cr}) < n < 1$ and one positive equilibrium for $n > 1$.

In the results above, the sign of the derivative $\nu'(p)$, or equivalently $n'(p)$, for small $p > 0$ determines the stability of the positive equilibrium pairs (n, x) near the bifurcation point $(1, 0)$. A natural question to ask is whether the derivative $n'(p)$ can tell us anything about the stability of positive equilibria in general.

Return, for the moment, to the general case $P(x) = T(x) + F(x)$. By definition $n(x)$ is the dominant eigenvalue of $F(x)(I - T(x))^{-1}$ with a nonnegative eigenvector $u(x) \geq 0$. $n(x)$ is also an eigenvalue of the matrix $(I - T(x))^{-1}F(x)$ with eigenvector $z(x) = (I - T(x))^{-1}u(x) > 0$. From the equation $F(x)z(x) = n(x)(I - T(x))z(x)$ one can derive a formula for the partial derivative $\partial_i n$. If this formula is evaluated at an equilibrium $x > 0$ (where $n(x) = 1$ and $z(x) = x$), one obtains

$$\partial_i n(I - T(x))x = (\partial_i P(x))\, x - (I - P(x))\, \partial_i z.$$

If both sides are multiplied by the left eigenvector $w^\tau(x)$ of $P(x)$ at equilibrium (for which $w^\tau(x) = w^\tau(x)P(x)$) we get the formula

$$\partial_i n(x) = \frac{w^\tau(x)\partial_i P(x)x}{w^\tau(x)F(x)x}. \tag{1.37}$$

Return now to the case where density dependence is through a dependence on a single weighted total population size p

$$P(p) = T(p) + F(p). \tag{1.38}$$

Then $n = n(p)$, $\partial_i n(x) = \omega_i n'(p)$, and $\partial_i P(x) = \omega_i P'(p)$ in (1.37). Choosing an i for which $\omega_i \neq 0$, we obtain from (1.37)

$$n'(p) = \frac{w^\tau(p)P'(p)x}{w^\tau(p)F(p)x}. \tag{1.39}$$

Can this derivative be related to the eigenvalues of the Jacobian $J(p)$?

As an example, consider the general nonlinear Usher matrix

$$P(p) = \begin{bmatrix} t_{11}(p) + f_{11}(p) & f_{12}(p) & \cdots & f_{1,m-1}(p) & f_{1m}(p) \\ t_{21}(p) & t_{22}(p) & \cdots & 0 & 0 \\ 0 & t_{32}(p) & \cdots & 0 & 0 \\ \vdots & \vdots & \ddots & \vdots & \vdots \\ 0 & 0 & \cdots & t_{m,m-1}(p) & t_{mm}(p) \end{bmatrix} \tag{1.40}$$

for which (see (1.15))

$$n(p) = \sum_{i=1}^{m} f_{1i}(p) \prod_{j=1}^{i} \frac{t_{j,j-1}(p)}{1 - t_{jj}(p)}.$$

A straightforward (but tedious) calculation shows (at equilibrium)

$$\det(I - J(p)) = -pn'(p) \prod_{i=1}^{m} (1 - t_{ii}(p)).$$

If $n'(p) > 0$, then the characteristic polynomial $\det(\zeta I - J(p))$ of the Jacobian is negative when evaluated at $\zeta = 1$. Since this polynomial tends to $+\infty$ for real $\zeta \to +\infty$, it follows that it must have a real root $\zeta > 1$. Thus, $n'(p) > 0$ implies that the positive equilibrium is unstable and $n'(p) \leq 0$ becomes a necessary condition for equilibrium stability. This necessary condition has been established for general matrix models of the form (1.38). The theorem below can be found in [442]. Let $c_{ii}(p)$ be the cofactor of the ith diagonal element in the matrix $I - P(p)$. (The strong dissipative condition $\sum_{i=1}^{m} t_{ij}(x) \leq 1$ is required in [442], but it is only needed to guarantee the existence of the net reproductive number. Conditions (1.34) are sufficient for this purpose and the proof in [442] is unchanged.)

THEOREM 1.2.10. *Consider the matrix equation* (1.33) *with a projection matrix of the form* (1.38) *under assumptions* (1.34). *Let $x > 0$ be a positive equilibrium with weighted total population size $p = |x| > 0$. Assume that $c_{ii}(p) > 0$ for all $i = 1, 2, \ldots, m$. Then $n'(p) > 0$ implies that x is unstable.*

The condition that $c_{ii}(p) > 0$ for all $i = 1, 2, \ldots, m$ can be shown to hold for general nonlinear Usher models (and hence for general nonlinear Leslie models). It can also be shown to hold for general matrix models of size $m = 2$ and 3; whether it holds for general matrix models of size $m \geq 4$ remains an open question [442].

While Theorem 1.2.10 shows that $n'(p) \leq 0$ is necessary for the stability of a positive equilibrium, neither this condition, nor the stronger condition $n'(p) < 0$, is sufficient to guarantee stability. The famous $m = 1$ dimensional Ricker map $x(t+1) = b\exp(-cx(t))x(t)$, for which $n(x) = b\exp(-cx)$, serves as a counterexample.

The geometric interpretation of Theorem 1.2.10 is that equilibria whose weighted total population size p lies on a decreasing branch of the graph of (1.35) in the (n, p) plane are unstable. See Fig. 1.7.

Finally, we point out that while we have used (1.32) to deduce properties of the bifurcating branch of equilibrium pairs when using the inherent net reproductive number n as the bifurcation parameter λ, this identity can also be used for the same purpose when other bifurcation parameters are used.

1.3 Some applications

The applications below illustrate the use of the results from the previous sections. The first application provides a general treatment of a large class of

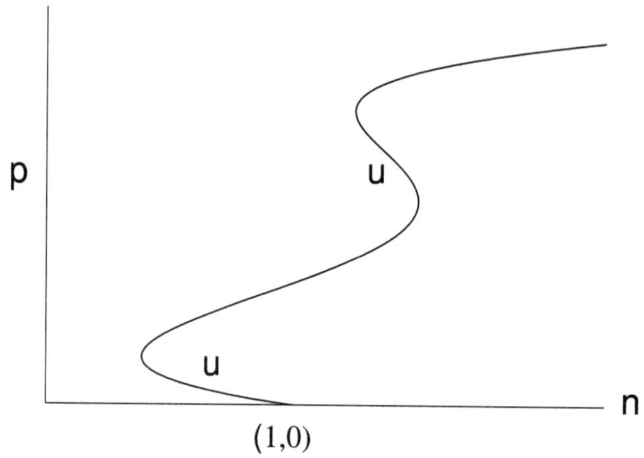

FIG. 1.7. *Under the conditions of Theorem 1.2.10 the decreasing portions of the equilibrium branch are unstable.*

nonlinear matrix models. The second application is to a specific model for the population dynamics of flour beetles and it illustrates a stable bifurcation from the extinction state (followed by either 2-cycle or discrete Hopf bifurcations). This matrix model is also used to illustrate how different choices of inherent parameters can be made for use as a bifurcation parameter. The third application involves a matrix model that has been used to study the dynamics of the controversial spotted owl of the Pacific northwest. This model illustrates an unstable bifurcation from the extinction state due to an Allee effect. The final example involves a size-structured model that has been used to conduct a theoretical study of the effects of competition on juvenile maturation size. This example illustrates a "nongeneric" bifurcation from the extinction state.

A general nonlinear Usher model. Consider a nonlinear Usher projection matrix (1.40) in which the density dependence is through a dependence on a weighted total population size $p = \sum_{i=1}^{m} \omega_i x_i$, $\omega_i \geq 0$, $\sum_{i=1}^{m} \omega_i \neq 0$. The net reproductive number is (see (1.15))

$$n(p) = \sum_{i=1}^{m} f_{1i}(p) \prod_{j=1}^{i} \frac{t_{j,j-1}(p)}{1 - t_{jj}(p)}.$$

Setting $f_{1i}(p) = n\varphi_{1i}(p)$, where the φ_{1i} are normalized so that

$$\sum_{i=1}^{m} \varphi_{1i}(0) \prod_{j=1}^{i} \frac{t_{j,j-1}(0)}{1 - t_{jj}(0)} = 1, \tag{1.41}$$

the number n is the inherent net reproductive number and the nonlinear Usher projection matrix has the form $P(p) = T(p) + n\Phi(p)$, where

$$T(p) = \begin{bmatrix} t_{11}(p) & 0 & \cdots & 0 & 0 \\ t_{21}(p) & t_{22}(p) & \cdots & 0 & 0 \\ 0 & t_{32}(p) & \cdots & 0 & 0 \\ \vdots & \vdots & \ddots & \vdots & \vdots \\ 0 & 0 & \cdots & t_{m,m-1}(p) & t_{mm}(p) \end{bmatrix},$$

$$\Phi(p) = \begin{bmatrix} \varphi_{11}(p) & \varphi_{12}(p) & \cdots & \varphi_{1m}(p) \\ 0 & 0 & \cdots & 0 \\ \vdots & \vdots & \ddots & \vdots \\ 0 & 0 & \cdots & 0 \end{bmatrix}.$$

If we assume, for an open interval $\Lambda \subseteq R^1$ containing the half line $[0, +\infty)$, the smoothness conditions

$$t_{ii} \in C^{k+1}\left(\Lambda, [0,1)\right), \quad t_{i,i-1} \in C^{k+1}\left(\Lambda, (0,1]\right), \quad \varphi_{1j} \in C^k\left(\Lambda, R^1_+\right)$$

for some integer $k \in I[0, +\infty)$ and in addition the inequalities

$$t_{i+1,i}(p) + t_{ii}(p) < 1, \quad \varphi_{1m}(p) > 0 \text{ for all } p \in R^1_+ ,$$

then the nonlinear Usher matrix satisfies the first four conditions (1.34(a)–(d)) in section 1.2.5. From

$$(I - T(p))^{-1} = \begin{bmatrix} \frac{1}{1-t_{11}(p)} & 0 & \cdots & 0 \\ \vdots & \vdots & \cdots & 0 \\ \prod_{j=1}^{i} \frac{t_{j,j-1}(p)}{1-t_{jj}(p)} & \prod_{j=2}^{i} \frac{t_{j,j-1}(p)}{1-t_{jj}(p)} & \cdots & 0 \\ \vdots & \vdots & \ddots & \vdots \\ \prod_{j=1}^{m} \frac{t_{j,j-1}(p)}{1-t_{jj}(p)} & \prod_{j=2}^{m} \frac{t_{j,j-1}(p)}{1-t_{jj}(p)} & \cdots & \frac{1}{1-t_{mm}(p)} \end{bmatrix}$$

we find

$$\Phi(p)(I - T(p))^{-1} = \begin{bmatrix} \sum_{i=1}^{m} \varphi_{1i}(p) \prod_{j=1}^{i} \frac{t_{j,j-1}(p)}{1-t_{jj}(p)} & \cdots & \varphi_{1m}(p) \frac{1}{1-t_{mm}(p)} \\ 0 & \cdots & 0 \\ \vdots & \ddots & \vdots \\ 0 & \cdots & 0 \end{bmatrix}$$

whose dominant eigenvalue and eigenvector are

$$\nu(p) = \sum_{i=1}^{m} \varphi_{1i}(p) \prod_{j=1}^{i} \frac{t_{j,j-1}(p)}{1 - t_{jj}(p)}, \tag{1.42}$$

$$u(p) = \begin{bmatrix} 1 \\ 0 \\ \vdots \\ 0 \end{bmatrix} \geq 0.$$

(For notational convenience we have defined $t_{10}(p) \equiv 1$.) Then for all $p \in R^1_+$

$$(I - T(p))^{-1} u(p) = \begin{bmatrix} \frac{1}{1-t_{11}(p)} \\ \vdots \\ \prod_{j=1}^{i} \frac{t_{j,j-1}(p)}{1-t_{jj}(p)} \\ \vdots \\ \prod_{j=1}^{m} \frac{t_{j,j-1}(p)}{1-t_{jj}(p)} \end{bmatrix} > 0,$$

and the last condition (1.34(e)) in section 1.2.5 is satisfied. Thus, all results in sections 1.2 and 1.2.5 hold for the general nonlinear Usher model.

As an illustrative example, consider a nonlinear Usher model with Ricker-type, exponential nonlinearities for fertilities and transition probabilities; specifically

$$\begin{aligned} &\varphi_{1i}(p) = \beta_i \exp(-c_i p), \quad \beta_i \geq 0, \quad c_i > 0, \\ &t_{i+1,i}(p) = \pi_i \exp(-d_i p), \quad 0 < \pi_i < 1, \quad d_i > 0, \\ &t_{ii}(p) = \pi_i (1 - \exp(-d_i p)). \end{aligned}$$

The normalized inherent fertilities β_i satisfy the normalization (1.41), namely,

$$\sum_{i=1}^{m} \beta_i \prod_{j=1}^{i} \pi_j = 1.$$

One interpretation of these submodels for the transition probabilities is that π_i is the probability that an individual in class i survives one unit of time and $\exp(-d_i p)$ is the probability that such a survivor moves into (e.g., "grows" into) the next class $i+1$. The number of newborns per individual of class i is β_i and the exponential $\exp(-c_i p)$ is the probability that a newborn from a parent of class i will survive to the next census time.

For this example, it is not difficult to see from (1.42) that $\nu'(p) < 0$ for all $p \geq 0$ and $\lim_{p\to+\infty} \nu(p) = 0$. Thus, from the results in section 1.2.5 we find that $x = 0$ loses stability as n increases through 1. For each $n > 1$ the nonlinear Usher matrix model is uniformly persistent with respect to the extinction equilibrium $x = 0$ (by Theorem 1.2.2) and there exists a unique positive equilibrium; these positive equilibria are (locally asymptotically) stable for $n \approx 1$ and are unbounded as $n \to +\infty$. Moreover, (1.22) holds and consequently $n < 1$ implies global extinction (i.e., $\lim_{t\to+\infty} |x(t)| = 0$ for all $x(0) \geq 0$).

In the preceding example only the facts that the submodels for the fertilities and transition probabilities tend monotonically to 0 as p increases without bound were used. Thus, other nonlinearities (or a mix of nonlinearities), such as $(1 + c_i p^{\alpha})^{-1}$, could be used with the same results.

The LPA flour beetle model. Flour beetles (*Tribolium*) live almost exclusively in stored grain products and have apparently done so for nearly as long as humans have stored and milled grains (e.g., evidence of their presence have been found in ancient Egyptian urns). Besides being a major agricultural pest, flour beetles are an excellent animal for laboratory experiments in population biology. For these reasons, flour beetles have been intensely studied for many decades and their physiological, behavioral, genetic, and life-cycle characteristics are very well understood. In [72], [73], [74], [116], [117], [131], [132], [133], [232] the following matrix model (called the *LPA model*)

$$\begin{aligned} L(t+1) &= bA(t)\exp\left(-c_{el}L(t) - c_{ea}A(t)\right), \\ P(t+1) &= \left(1-\mu_l\right)L(t), \\ A(t+1) &= P(t)\exp\left(-c_{pa}A(t)\right) + \left(1-\mu_a\right)A(t) \end{aligned} \tag{1.43}$$

is used to describe the population dynamics of laboratory cultures of flour beetles. In this model $L(t)$ is the number of larvae, $P(t)$ is the number of pupae (but also including nonfeeding larvae and young nonreproducing adults), and $A(t)$ is the number of reproducing adults at time t. The time unit (census interval) is two weeks. The nonlinear interactions in this model, as described by the exponential terms, are attributed to cannibalism between certain life-cycle stages, and the parameters $c_{el} > 0$, $c_{ea} > 0$, and $c_{pa} > 0$ are called the "cannibalism coefficients." Thus, the inherent number of larvae produced per adult per unit of time, $b > 0$, is reduced by the fraction $\exp(-c_{el}L - c_{ea}A)$ due to larval and adult cannibalism of eggs. Pupal mortality is due only to cannibalism by adults. Larvae and adults have constant death rates $0 < \mu_l < 1$ and $0 < \mu_a < 1$, respectively. Cannibalism in flour beetles occurs from random encounters and is therefore well modeled by exponential type nonlinearities (see section 1.1.3).

The LPA model (1.43) has the form of a nonlinear Leslie matrix model with

$$P(x) = \begin{bmatrix} 0 & 0 & b\exp\left(-c_{el}L - c_{ea}A\right) \\ 1-\mu_l & 0 & 0 \\ 0 & \exp\left(-c_{pa}A\right) & 1-\mu_a \end{bmatrix}, \quad x = \begin{bmatrix} L \\ P \\ A \end{bmatrix},$$

and inherent net reproductive number

$$n = b\frac{1-\mu_l}{\mu_a}.$$

It can be put in the form (1.33) with transition matrix

$$T(x) = \begin{bmatrix} 0 & 0 & 0 \\ 1-\mu_l & 0 & 0 \\ 0 & \exp\left(-c_{pa}A\right) & 1-\mu_a \end{bmatrix}$$

and normalized fertility matrix

$$\Phi(x) = \begin{bmatrix} 0 & 0 & \frac{\mu_a}{1-\mu_l}\exp\left(-c_{el}L - c_{ea}A\right) \\ 0 & 0 & 0 \\ 0 & 0 & 0 \end{bmatrix}.$$

Note that

$$\Phi(x)\left(I-T(x)\right)^{-1}=\exp\left(-c_{el}L-\left(c_{ea}+c_{pa}\right)A\right)\begin{bmatrix}1 & \frac{1}{1-\mu_l} & \frac{1}{1-\mu_l}e^{c_{pa}A}\\ 0 & 0 & 0\\ 0 & 0 & 0\end{bmatrix}$$

so that

$$\nu(x)=\exp\left(-c_{el}L-\left(c_{ea}+c_{pa}\right)A\right),\quad u(x)=\begin{bmatrix}1\\0\\0\end{bmatrix}\geq 0,$$

$$\left(I-T(x)\right)^{-1}u(x)=\begin{bmatrix}1\\ 1-\mu_l\\ \frac{1-\mu_l}{\mu_a}e^{-c_{pa}A}\end{bmatrix}>0.$$

Thus, all conditions in (1.34) are satisfied (for all $k\geq 0$).

Since $P(x)\leq P(0)$ for all $x\in R_+^3$, we find from Theorems 1.1.3 and 1.2.1 that $n<1$ implies that the origin $x=0$ is globally attracting in R_+^3 and the model predicts asymptotic extinction. From these same theorems we have that the origin is unstable for $n>1$. Unfortunately, the transition matrix associated with $P(x)$ does not satisfy the condition (1.24) and therefore we cannot apply Theorem 1.2.2 directly to show that the LPA model (1.43) is dissipative. However, this can be shown as follows. From the first and second equations in (1.43) we have the inequalities

$$\begin{aligned}0 &\leq L(t+1)\leq b\frac{1}{c_{ea}e}=n\frac{\mu_a}{1-\mu_l}\frac{1}{c_{ea}e},\\ 0 &\leq P(t+2)\leq b\left(1-\mu_l\right)\frac{1}{c_{ea}e}=n\mu_a\frac{1}{c_{ea}e}\end{aligned}$$

which show that after two time steps the L and P components of all solutions are bounded by common bounds. From the third equation in (1.43) we have

$$0\leq A(t+3)\leq b\left(1-\mu_l\right)\frac{1}{c_{ea}e}+\left(1-\mu_a\right)A(t+2)$$

from which we conclude, using an induction argument as in the proof of Theorem 1.2.2,

$$0\leq A(t)\leq 2b\frac{1-\mu_l}{\mu_a}\frac{1}{c_{ea}e}=2n\frac{1}{c_{ea}e}$$

for sufficiently large $t\geq 0$; i.e., (1.43) is dissipative. Uniform persistence (with respect to the origin) follows from Theorem 1.2.1 for $n>1$. Theorems 1.2.7 and 1.2.6 guarantee that the bifurcation, from the origin at $n=1$, of a global continuum of positive equilibrium pairs which has a positive spectrum, is unbounded

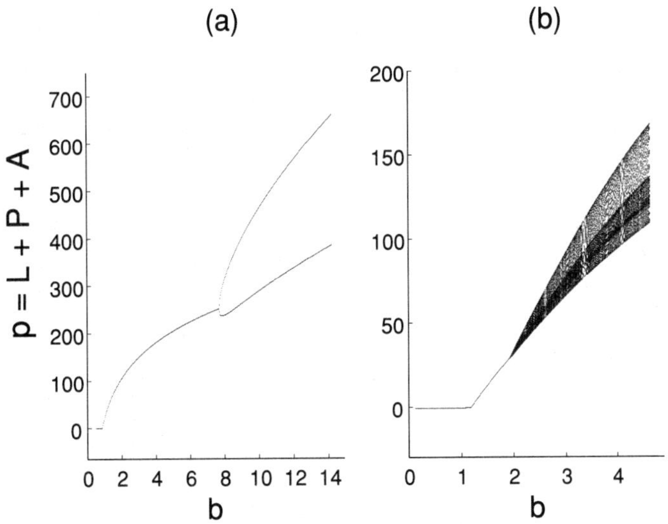

FIG. 1.8. *Two bifurcation diagrams for the flour beetle model* (1.43) *show the attractor, represented by total population size* $p = L + P + A$*, plotted against the parameter* b*. In* (a), $c_{ea} = c_{el} = 0.01$, $c_{pa} = 0.05$, $\mu_l = 0.2$, *and* $\mu_a = 0.9$, *and in* (b), $c_{ea} = c_{el} = c_{pa} = 0.01$, $\mu_l = 0.5$, *and* $\mu_a = 0.20$. *To eliminate transients,* 1000 *iterations were performed before* 100 *values of* p *were plotted.*

in $R^1_+ \times R^3_+$ and is (locally asymptotically) stable near the bifurcation point. The inequalities above (which are valid for all solutions, including equilibrium solutions) show that the magnitude of equilibria are bounded above by a multiple of n, and from this it follows that the spectrum of the continuum cannot be bounded (for if it were, then the global continuum would be bounded). It is not difficult to show that the positive equilibrium for $n > 1$ is unique.

The equilibrium stability for n near 1 does not in general persist for large n. The bifurcation diagrams in Fig. 1.8 show that both 2-cycle bifurcations and discrete Hopf (invariant loop) bifurcations are possible for the LPA model (1.43), depending on parameter values. Fig. 1.8(a) shows a stable 2-cycle bifurcation occurring at a critical value of $n > 1$, where the positive equilibrium loses stability (because an eigenvalue of the Jacobian decreases through -1 as n increases), and Fig. 1.8(b) shows the bifurcation of an invariant loop (because a complex conjugate pair of eigenvalues leaves the complex unit circle).

In the experiments described in [72], [133] the adults' death rate μ_a was manipulated. We can also study (1.43) using the inherent death rate $\lambda = \mu_a$ as a bifurcation parameter. To do this we write the inherent projection

$$P(0) = \begin{bmatrix} 0 & 0 & b \\ 1 - \mu_l & 0 & 0 \\ 0 & 1 & 1 - \mu_a \end{bmatrix}$$

as $P(0) = A + \lambda B$, where

$$A = \begin{bmatrix} 0 & 0 & b \\ 1-\mu_l & 0 & 0 \\ 0 & 1 & 1 \end{bmatrix}, \quad B = \begin{bmatrix} 0 & 0 & 0 \\ 0 & 0 & 0 \\ 0 & 0 & -1 \end{bmatrix}.$$

The only critical value of $A + \lambda B$ is $\lambda_0 = (1-\mu_l)\, b$ and it is simple and strictly dominant. The right and left eigenvectors of $A + \lambda_0 B$ associated with the eigenvalue 1 are

$$v = \begin{bmatrix} b \\ (1-\mu_l)\, b \\ 1 \end{bmatrix} > 0, \quad w = \begin{bmatrix} 1-\mu_l \\ 1 \\ 1 \end{bmatrix} > 0$$

from which we calculate $w^\tau Bv = -1 < 0$.

Theorem 1.2.3 implies that the extinction equilibrium $x = 0$ loses stability as $\lambda = \mu_a$ *decreases* through λ_0, and Theorem 1.2.4 implies the existence of a bifurcating continuum C^+ of equilibrium pairs that satisfies the alternatives of that theorem. However, the uniqueness of the critical value λ_0 rules out alternative (b). Moreover, it is straightforward to show, from the equilibrium equations, that a nonnegative equilibrium must be positive; this rules out alternative (c). The remaining alternative, (a), implies that the continuum C^+ is unbounded in $R^1 \times R^3_+$ and consists of positive equilibrium pairs (λ, x) (except for the bifurcation point $(\lambda_0, 0)$). The identity (1.32), which must hold for all equilibrium pairs from C^+, is

$$b\frac{1-\mu_l}{\lambda} \exp\left(-c_{el}L - (c_{ea} + c_{pa})\, A\right) = 1. \tag{1.44}$$

Thus, the spectrum $\sigma\,(C^+)$ cannot be unbounded. The unboundedness of C^+ then implies that the range $\rho\,(C^+)$ is unbounded. By (1.44) $\lambda = 0$ is in the closure of $\sigma\,(C^+)$. Since (1.44) implies that $\lambda < \lambda_0$, we deduce that $\sigma\,(C^+)$ is identical with the interval $0 < \lambda \leq \lambda_0$ and the bifurcation is to the left.

Furthermore, the formulas in Theorem 1.2.5 yield

$$[d_{ij}]\, v = -b \begin{bmatrix} bc_{el} + c_{ea} \\ 0 \\ c_{pa}\,(1-\mu_l) \end{bmatrix}, \quad w^\tau\, [d_{ij}]\, v = -b(1-\mu_l)\,(bc_{el} + c_{ea} + c_{pa}) < 0,$$

and hence

$$\lambda_1 = -\frac{w^\tau\, [d_{ij}]\, v}{w^\tau Bv} = -b(1-\mu_l)\,(bc_{el} + c_{ea} + c_{pa}) < 0.$$

Theorems 1.2.5 and 1.2.6 imply that the bifurcation is stable.

As we have noted above, the spectrum of $\lambda = \mu_a$ values associated with positive equilibria from C^+ is the interval $0 < \lambda < \lambda_0$. Thus, either $\lambda_0 \geq 1$ and the extinction equilibrium is unstable for all $\lambda = \mu_a$ of biological interest $(0 < \mu_a < 1)$ or $\lambda_0 < 1$ and the bifurcation occurs at a biologically meaningful value of the adult death rate μ_a. See Fig. 1.9.

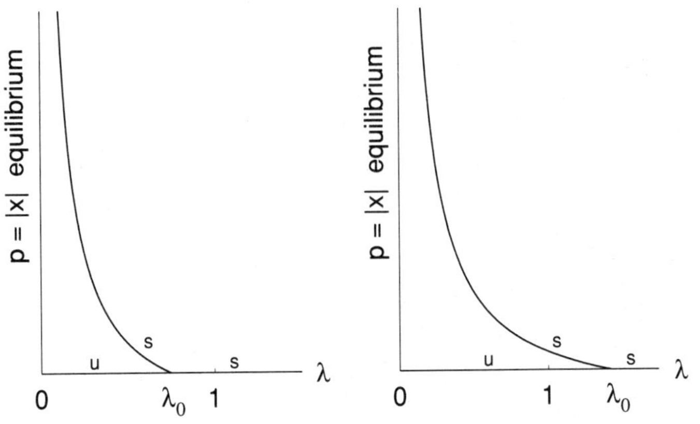

FIG. 1.9. *The two possible bifurcation diagrams for the beetle model* (1.43) *using the adult death rate* $\lambda = \mu_a$ *as the bifurcation parameter are shown. On the left* $\lambda_0 = (1 - \mu_l)\, b < 1$, *while on the right* $\lambda_0 > 1$. *The local stability of the positive equilibria near the bifurcation point* $(\lambda_0, 0)$ *may not persist for all values of* $\lambda = \mu_a < 1$.

A spotted owl model. The northern spotted owl is a long-lived, territorial predator that requires large tracts of old growth forest in which to live. Interest in the population dynamics of the spotted owl was intensified during the 1980s by an environmental controversy concerning the logging of old growth forests in the Pacific northwest. A matrix model for these dynamics was introduced and studied in [272], [275] (also see [209]). This model describes the dynamics of the $m = 2$ variables

$$\begin{aligned} S(t) &= \text{ number of single male owls at time } t, \\ C(t) &= \text{ number of paired adults (reproductive couples) at time } t \end{aligned}$$

by a matrix equation with the projection matrix

$$P(x) = \begin{bmatrix} s_s\,(1 - M(S)) & \frac{1}{2} f s_j D(S,C) + p_f \\ 2s_s M(S) & p_s \end{bmatrix}, \quad x = \begin{bmatrix} S \\ C \end{bmatrix}.$$

The unit of time is one year. This model is based upon the life cycle of the spotted owl. After a year as juveniles, males establish territories and attract breeding females. Males are either single or paired with a female (monogamously). A juvenile must find a territory by the beginning of the second year (or it will die or leave the area). Timber harvesting fragments the habitat into small patches, which influences the population dynamics through effects on territory availability (nesting sites) and mating success, dispersal, etc. In the projection matrix above

$$M(S) \doteq 1 - \left(1 - 2\tfrac{S}{T}\right)^p = \text{ probability of a female finding a mate,}$$

$$D(C,S) \doteq 1 - \left(1 - \tfrac{U-C-S}{T}\right)^q = \text{ probability a juvenile survives dispersal.}$$

The model parameters (taken here to be constants) are

$s_s =$ fraction of single males surviving one year,
$s_j =$ fraction of juveniles surviving one year (to become single adults),
$f =$ number of offspring per adult pair per year,
$p_f =$ probability that only the female of a pair dies in a year,
$p_s =$ probability both individuals in a pair survive one year,
$q =$ unoccupied site search efficiency by single males (number of search attempts in a year before dying),
$p =$ unmated female search efficiency (number of search attempts in a year before dying or leaving the area),
$T =$ total number of sites in the system,
$U =$ number of suitable sites in the system.

The inherent projection matrix

$$P(0) = \begin{bmatrix} s_s & p_f + \frac{1}{2} f s_j \left[1 - \left(1 - \frac{U}{T}\right)^q\right] \\ 0 & p_s \end{bmatrix}$$

has eigenvalues $s_s < 1$ and $p_s < 1$, and therefore the extinction equilibrium is (locally asymptotically) stable for all model parameter values. Thus, in this model small populations will go extinct. Does the model predict that all populations will go extinct? Or are there stable positive equilibria?

If we want to study this question using the bifurcation theory from the preceding sections, we need an inherent parameter which, when varied, can lead to destabilization of the extinction equilibrium $x = 0$. The only choices are s_s or p_s. If, for example, p_s is mathematically allowed to increase through 1, $x = 0$ will become unstable. Of course, values of $p_s > 1$ are not biologically meaningful in this model, so ultimately we must restrict attention to $p_s \leq 1$.

Let $\lambda = p_s$, and write $P(0) = A + \lambda B$, where

$$A = \begin{bmatrix} s_s & p_f + \frac{1}{2} f s_j \left[1 - \left(\frac{T-U}{T}\right)^q\right] \\ 0 & 0 \end{bmatrix}, \quad B = \begin{bmatrix} 0 & 0 \\ 0 & 1 \end{bmatrix}.$$

The only critical value of $A + \lambda B$ is $\lambda_0 = 1$. From the eigenvectors

$$v = \begin{bmatrix} 2p_f + f s_j \left[1 - \left(\frac{T-U}{T}\right)^q\right] \\ 2(1 - s_s) \end{bmatrix} > 0, \quad w = \begin{bmatrix} 0 \\ 1 \end{bmatrix} \geq 0$$

we find $w^\tau B v = 2(1 - s_s) > 0$. Theorem 1.2.4 implies the existence of a bifurcating continuum C^+ of equilibrium pairs that satisfies the alternatives of that theorem. However, the uniqueness of the critical value λ_0 rules out alternative (b). We will rule out alternative (a) below. Before doing that, however, we note that the bifurcation is to the left and unstable. This follows from a calculation that shows

$$\lambda_1 = -\frac{2}{1 - s_s} s_s \frac{p}{T} \left(2p_f + f s_j \left[1 - \left(\frac{T-U}{T}\right)^q\right]\right)^2 < 0.$$

Theorems 1.2.5 and 1.2.6 imply that the bifurcation is to the left and unstable.

The equilibrium equations are

$$\begin{aligned} S &= s_s S\left(1-2\frac{S}{T}\right)^p + \frac{1}{2} f s_j C\left[1-\left(1-\frac{U-C-S}{T}\right)^q\right] + p_f C, \\ C &= 2s_s S\left[1-\left(1-2\frac{S}{T}\right)^p\right] + \lambda C. \end{aligned}$$

Note that the second equation implies that $\lambda < 1$ for any positive equilibrium. Also, from these equations one can deduce that the only nonextinction equilibrium pair $(\lambda^*, x^*) \in R^1 \times \partial R^2_+$, $0 \le x^* \ne 0$, available for alternative (c) in Theorem 1.2.4, is

(1.45a)
$$\lambda^* = 1, \quad x^* = \begin{bmatrix} 0 \\ C^* \end{bmatrix},$$
$$C^* \doteq \left(\tfrac{U}{T} + \left(\tfrac{f s_j + 2p_f}{f s_j}\right)^{\frac{1}{q}} - 1\right) T > 0.$$

A study of the equilibrium equations will show that alternative (a) of Theorem 1.2.4 (that C^+ is unbounded in the positive quadrant) is ruled out. This can be done by a contradiction argument, ruling out the possibility of the unboundedness of the spectrum and of the components S and C. We conclude from Theorem 1.2.4(c) that the continuum C^+ contains the point (λ^*, C^*) given by (1.45a).

We have shown that there exists a continuum of positive equilibrium pairs (λ, x) with $\lambda < 1$ which connects the equilibrium pairs $(1, 0)$ and $(1, x^*)$. It follows that the continuum must "turn around" at some point $\lambda_{cr} < 1$, where a saddle-node bifurcation occurs, and there exist at least two positive equilibria for λ on the interval $\lambda_{cr} < \lambda < 1$; see Fig. 1.10. Although we have no proof of this (except for the "lower" branch near $\lambda = 1$), we suspect that the "smaller" equilibria on the "lower" branch in Fig. 1.10 are unstable while the "larger" equilibria on the upper branch are stable. Computer simulations bear this out.

The biological conclusion is that while sufficiently large populations of the spotted owl are viable, small populations are in danger of extinction. This threshold phenomenon (its estimation, its dependence on model parameters, etc.) is the focus of the study in [272].

A juvenile growth model. Intraspecific competition between juveniles and adults has received considerable attention in the literature. One question of interest concerns the "stabilizing" or "destabilizing" effects of such an interaction. Generally, competition between juveniles and adults is considered destabilizing, although there can be exceptions and such an assertion depends on exactly what is meant by "destabilization." Studies of this question using discrete models can be found in [97], [112], [151], [152], [153], [299], and [110]. (Studies using continuous models can be found in [97], [113], [312], and [402].)

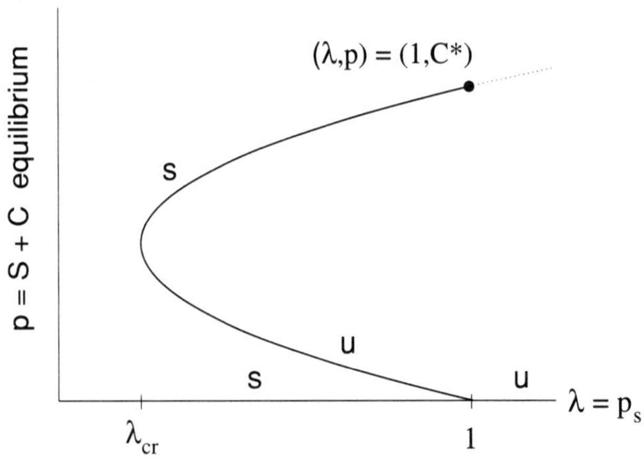

FIG. 1.10. *The unstable bifurcation in the spotted owl model at $\lambda = p_s = 1$ has a saddle-node bifurcation at λ_{cr}. The dotted lines at $(\lambda, p) = (1, C^*)$ represent equilibria that have left the positive quadrant.*

The nature of juvenile versus adult competition can be quite diverse, depending on details of the life cycle of the species. Most mathematical models in the literature utilize age structure. However, it is often pointed out that body size, rather than chronological age, is more often the determining factor in an individual's interaction with its physical and biological environment and hence in the effects that competition has on vital rates; see [154], [435]. A size-structured model for juvenile versus adult competition is studied in [153]; this model is not analytically tractable, however, and its study relied on computer simulations. Here we will consider a simpler, but more tractable, size-structured model introduced in [112]. This model focuses on how competition with adults effects the size of a juvenile at maturation and hence its adult fertility.

The fertility and transition matrices

$$F = \begin{bmatrix} 0 & nf_1 & nf_2 \\ 0 & 0 & 0 \\ 0 & 0 & 0 \end{bmatrix}, \qquad T = \begin{bmatrix} 0 & 0 & 0 \\ s & 0 & 0 \\ 1-s & 0 & 0 \end{bmatrix} \tag{1.46}$$

describe a population with a stage class distribution vector

$$x = \begin{bmatrix} J \\ A_1 \\ A_2 \end{bmatrix},$$

where J is the number (or density) of a juvenile class and A_1 and A_2 are the numbers (or density) of individuals in two adult classes whose fertility rates are nf_1 and nf_2, respectively. A fraction s of surviving juveniles become smaller adults after one unit of time, the remaining fraction $1 - s$ become larger adults.

Competition for resources is assumed to affect a juvenile's size at maturation. This is modeled by assuming $s = s(p)$ is a function of the weighted total population size

$$p \doteq J + \sigma\left(\beta_1 A_1 + \beta_2 A_2\right), \quad \beta_i \geq 0, \quad \sigma > 0. \tag{1.47}$$

The weights β_i measure the relative strength of the competition from the two adult classes as experienced by juveniles. Increased competition is assumed to result in less juvenile growth and hence smaller adults, so $s(p)$ is assumed to be an increasing function of p, as follows:

$$s \in C^1\left(R^1, [0,1]\right), \quad s'(p) > 0, \quad \lim_{p\to+\infty} s(p) = 1,$$

the last condition implying that as population density increases without bound the fraction of juveniles growing to become large adults drops to 0. The coefficients β_i satisfy the normalization

$$\beta_1 s(0) + \beta_1\left(1 - s(0)\right) = 1,$$

and σ measures the strength of adult competition experienced by a juvenile relative to competition from other juveniles.

The two adult classes are distinguished by the size of the individuals in them, the individuals in the first class A_1 being the smaller. The larger adults are assumed to be more fertile, and therefore $0 \leq f_1 < f_2$. The coefficients f_1 and f_2 satisfy the normalization

$$f_1 s(0) + f_1\left(1 - s(0)\right) = 1$$

so that n is the inherent net reproductive number. Adults survive only one time unit (or at least remain fertile only one time unit), and therefore this model is appropriate only for semelparous populations.

From

$$F(p)(I - T(p))^{-1} = \begin{bmatrix} nf_1 s(p) + nf_2\left(1 - s(p)\right) & nf_1 & nf_2 \\ 0 & 0 & 0 \\ 0 & 0 & 0 \end{bmatrix}$$

we obtain the net reproductive number $n(p) = n\nu(p)$, where

$$\nu(p) \doteq f_1 s(p) + f_2\left(1 - s(p)\right).$$

We can use the inherent net reproductive number n as a bifurcation parameter and apply Theorems 1.2.4 and 1.2.5 to obtain a continuum C^+ of equilibrium pairs that bifurcates from the extinction equilibrium. From $n\nu(p) = 1$ and $\nu'(p) = (f_1 - f_2)s'(p) < 0$ we deduce that the bifurcation is to the right. We would expect, from Theorem 1.2.6, that the bifurcation is stable. This theorem, however, cannot be applied since $n = 1$ is not a *strictly dominant* critical value (because the inherent projection matrix has eigenvalues ± 1 when $n = 1$).

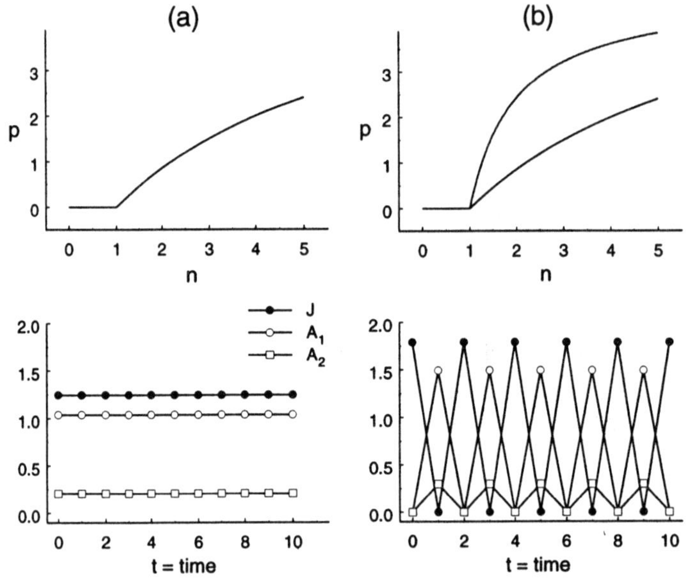

FIG. 1.11. *In the upper two graphs, bifurcation diagrams for the juvenile growth model* (1.46)–(1.47) *with* $s(p) = 1 - \exp(-p)$ *are shown for two different values of the competition coefficient* σ, *namely,* $\sigma = 0.5$ *in* (a) *and* $\sigma = 5.0$ *in* (b). *The values of* p *on the attractors are plotted as a function of the inherent net reproductive number* n. *Other parameters are* $f_1 = 0.1$, $f_2 = 1.0$, $\beta_1 = 0.25$, $\beta_2 = 1.0$. *In* (a) *the bifurcating equilibria are stable while in* (b) *the bifurcating* 2-*cycles are stable. The times series plots of these attractors are shown for* $n = 4$ *in the lower graphs. In* (b) *note that the resulting synchronous* 2-*cycle is such that juveniles and adults do not appear at the same time.*

In fact, it is shown in [112] that an additional continuum consisting of 2-cycles also bifurcates to the right from the extinction equilibrium at $n = 1$. (The existence of this continuum is suggested by the eigenvalue -1 at $n = 1$.) These 2-cycles are "synchronous" in the sense that juvenile and adult numbers periodically alternate between a positive value and 0 in an out-of-phase manner. This means that in these 2-cycles adults and juveniles are never present at the same time (thus avoiding direct competition!). It is shown in [112] that only one of the two bifurcating branches can be stable. Specifically, the positive equilibrium branch is stable near $n = 1$ if $\sigma < 1$ and the synchronous 2-cycles are stable near $n = 1$ if $\sigma > 1$. See Fig. 1.11. For large values of n the situation is more complicated, however; see [112].

1.4 A case study

Can mathematical models be used to describe and predict the dynamics of a real biological population? Can model predicted dynamics be observed in population data? Can predicted bifurcations in dynamics be observed in a population when physical or biological parameters are changed? Given the complexities of the biological world it is reasonable to begin by asking these questions in a controlled laboratory setting. In this section we give a brief description of a project that addresses these questions by means of controlled laboratory experiments and a relatively simple matrix model.

Mathematical models are always simplifications, of course, so that an important first step in building a model is to determine the dominant biological and physiological mechanisms that account for the dynamics of the population of interest under the relevant environmental circumstances. These mechanisms are to be incorporated into the model, while other mechanisms considered to be of less importance are ignored. The next crucial issue is that of connecting the resulting mathematical model with data. That is to say, how are model parameters to be assigned numerical values (i.e., how is the model "calibrated")? How is the resulting model "validated"; i.e., how are the model and the data used in the parameterization to be compared and a decision made as to how well the model "fits" the data? How are the model and data *not* used in the parameterization to be compared and the model's predictive capability to be judged? A thorough study of these issues fall outside the scope of this monograph. In section 1.4.1 a very brief introduction to one approach to these problems is given.

In section 1.4.2 a project involving controlled laboratory experiments is described in which a matrix model and its predicted bifurcations constitute the foundation on which the experiment protocol is based. This case study shows convincingly that a matrix model can describe and predict the asymptotic dynamics of a biological population (in this case, flour beetles). The model used in this study has been successful in other contexts as well. For example, model predicted unstable equilibria, together with their stable manifolds, accurately describe and explain certain transient behavior observed in laboratory data [117]. In section 1.4.3 a modification of the model is used to explain an unexpected "resonance" phenomenon observed in experiments involving periodically fluctuating habitats (and to predict some unintuitive dynamics that were subsequently confirmed by means of further experiments) [262].

1.4.1 Model parameterization and validation. One way to obtain a numerical value for a model parameter is directly from experimental measurements for that parameter. For example, if a constant survival fraction t_{ii} appears in a matrix model and if one has census counts of both living and dead i class individuals after one time unit, then one can estimate this fraction from this census data. Procedures for obtaining parameter estimates in this way, from "life history" data, are discussed for linear matrix models in [55].

Another way to obtain parameter values in a model is to control that parameter by environmental or physical manipulations, assuming this can be done.

Often (if not usually) neither of these options is available for some or all of the model parameters. In this case, one can attempt to obtain parameter estimates from observed time series data, which we assume are observed values of the state variables $x_i(t)$ in a matrix model at a finite number of census times $t \in I[0, q]$. Given parameter values the matrix equation provides, at each time t, a (deterministic) prediction for the data at the subsequent time $t + 1$. One is interested, of course, in the differences between the model prediction and the actual observed data at the subsequent time $t + 1$. These "residuals" can be used both to obtain parameter estimates (by choosing parameter values that minimize the residuals) and for model performance (by a study of their statistical properties).

However, biological data tend to be "noisy," i.e., to have "random" fluctuations not accounted for in the deterministic model. In order to relate data to a model adequately one needs to account for this stochasticity. This requires a "stochastic model" in which appropriate random variables have been included which reflect the type of noise present and its effect on the components of the model. The resulting stochastic model (i.e., the deterministic model "skeleton" and the hypothesized noise structure) provides a "likelihood function" for one step predictions. The stochastic model can be viewed as a "testable hypothesis" and its one-step residuals provide a means for its parameterization and model validation.

Although this monograph deals only with deterministic models, we will illustrate these general remarks with some simple examples in order to provide an understanding of the case study in section 1.4.2. For more details of these issues see [132], [116].

Consider a one-dimensional ($m = 1$) map

$$x(t + 1) = f(x(t), \theta_1, \ldots, \theta_p) \tag{1.48}$$

with p parameters θ_i. For given values of θ_i this map makes the deterministic prediction that, given a population of size $x(t)$ at time t, the total population size at time $t + 1$ will be $f(x(t), \theta_1, \ldots, \theta_p)$. In general, the actual total population size at $t + 1$ will deviate from this prediction in an unpredictable way and we wish to modify this map to reflect this stochasticity.

Ecologists draw a distinction between two broad types of stochasticity. "Environmental" stochasticity affect many individuals in the population and are usually due to extrinsic factors. "Demographic" stochasticity involves chance variations among individuals. Environmental stochasticity, which typically predominates at high population levels, is additive on the logarithmic scale [131].

As an example, consider noise added on the logarithmic scale to the following model (1.48):

$$x(t + 1) = f(x(t), \theta_1, \ldots, \theta_p)e^{E_t}.$$

Here E_t is a normal random variable with mean 0 and variance v. We assume that E_0, $E_1, \ldots$ are uncorrelated. A stochastic model of this form is a type of multivariate, nonlinear, autoregressive model [403]. The nonlinear deterministic skeleton (1.48) is preserved on the logarithmic scale for the expectations of the random variable $x(t)$ in the sense that

$$E\left(\ln x(t + 1) | \, x(t) = y_t\right) = \ln\left(f(y_t, \theta_1, \ldots, \theta_p)\right).$$

Suppose that a time series of $q+1$ data points

$$y_0,\ y_1,\ y_2, \dots,\ y_q$$

is given. A "likelihood function" L gives the probability that the observed data would result from the proposed stochastic mechanism relative to all other possible outcomes [132]. The data y_t is a realization of the random variable $x(t)$. On the log scale, $w_t = \ln y_t$ is a realization of the random variable $\ln x(t)$. The likelihood function L is

$$L(\theta_1, \dots, \theta_p, v) = \prod_{t=1}^{q} p(w_t \mid w_{t-1}),$$

where $p(w_t \mid w_{t-1})$ is the joint probability distribution function (pdf) that w_t occurs given that w_{t-1} occurs. This is a normal pdf with mean $\ln f(y_{t-1}, \theta_1, \dots, \theta_p)$ and variance v. Thus,

$$p(w_t \mid w_{t-1}) = \frac{1}{\sqrt{2\pi v}} \exp\left(-\frac{1}{2v}\left(w_t - \ln f(y_{t-1}, \theta_1, \dots, \theta_p)\right)^2\right)$$

and

$$L(\theta_1, \dots, \theta_p, v) = \prod_{t=1}^{q} \frac{1}{\sqrt{2\pi v}} \exp\left(-\frac{1}{2v}\left(w_t - \ln f(y_{t-1}, \theta_1, \dots, \theta_p)\right)^2\right).$$

The *maximum likelihood parameter estimates* are those values of the parameters $\theta_1, \dots, \theta_p, v$ that maximize $L(\theta_1, \dots, \theta_p, v)$, or equivalently that maximize $l(\theta_1, \dots, \theta_p, v) \doteq \ln\left(L(\theta_1, \dots, \theta_p, v)\right)$. A calculation shows

$$l(\theta_1, \dots, \theta_p, v) = -\frac{q}{2}\ln(2\pi) - \frac{q}{2}\ln v - \frac{1}{2v}\sum_{t=1}^{q} r_t^2(\theta_1, \dots, \theta_p), \tag{1.49}$$

where

$$r_t(\theta_1, \dots, \theta_p) \doteq \ln y_t - \ln f(y_{t-1}, \theta_1, \dots, \theta_p) = \ln\left(\frac{y_t}{f(y_{t-1}, \theta_1, \dots, \theta_p)}\right)$$

are the "log-residuals." The critical points $(\theta_1, \dots, \theta_p, v)$ of l are zeros of the derivatives

$$\begin{aligned} \partial_{\theta_i} l &= -\frac{1}{v}\sum_{t=1}^{q} r_t(\theta_1, \dots, \theta_p)\partial_{\theta_i} r_t(\theta_1, \dots, \theta_p), \\ \partial_v l &= -\frac{q}{2}\frac{1}{v} + \frac{1}{2v^2}\sum_{t=1}^{q} r_t^2(\theta_1, \dots, \theta_p), \end{aligned}$$

i.e., are roots of the uncoupled equations

$$\sum_{t=1}^{q} r_t(\theta_1, \dots, \theta_p)\frac{\partial_{\theta_i} f(y_{t-1}, \theta_1, \dots, \theta_p)}{f(y_{t-1}, \theta_1, \dots, \theta_p)} = 0, \tag{1.50}$$

(1.51)
$$v = \frac{1}{q}\sum_{t=1}^{q} r_t^2(\theta_1, \dots, \theta_p).$$

The first equation represents p equations for the p parameters $\theta_1, \dots, \theta_p$ whose solution gives the maximum likelihood estimates for $\theta_1, \dots, \theta_p$. The second equation then gives the maximum likelihood estimate for the variance v in the stochastic version of the model. In general, the equations (1.50) must be solved numerically.

As an illustration consider the following stochastic version of the Ricker equation:

(1.52)
$$x(t+1) = bx(t)e^{-cx(t)}e^{E_t}.$$

Here there are two parameters, $\theta_1 = b$ and $\theta_2 = c$, in the deterministic skeleton. For this model the equations (1.50), namely,

$$\begin{aligned} q \ln b + \left(-\sum_{t=1}^{q} y_{t-1}\right) c &= \sum_{t=1}^{q} \ln\left(\frac{y_t}{y_{t-1}}\right), \\ \left(\sum_{t=1}^{q} y_{t-1}\right) \ln b + \left(-\sum_{t=1}^{q} y_{t-1}^2\right) c &= \sum_{t=1}^{q} y_{t-1} \ln\left(\frac{y_t}{y_{t-1}}\right), \end{aligned}$$

are linear in $\ln b$ and c with solutions given by

$$\ln b = \frac{s_1 s_4 - s_3 s_2}{s_1^2 - q s_2}, \qquad c = \frac{q s_4 - s_1 s_3}{s_1^2 - q s_2},$$

where

$$\begin{aligned} s_1 &\doteq \sum_{t=1}^{q} y_{t-1}, \qquad s_2 \doteq \sum_{t=1}^{q} y_{t-1}^2, \\ s_3 &\doteq \sum_{t=1}^{q} \ln\left(\frac{y_t}{y_{t-1}}\right), \qquad s_4 \doteq \sum_{t=1}^{q} y_{t-1} \ln\left(\frac{y_t}{y_{t-1}}\right). \end{aligned}$$

From (1.51)

$$v = \frac{1}{q}\left(q \ln^2 b + (-2c \ln b)\, s_1 + c^2 s_2 + (-2 \ln b)\, s_3 + 2c s_4 + s_5\right),$$

where

$$s_5 \doteq \sum_{t=1}^{q} \ln^2\left(\frac{y_t}{y_{t-1}}\right).$$

As an illustration of these formulas, we generate "simulation data" y_t from the deterministic ($v = 0$) Ricker map with $b = 10$, $c = 1/100$ and $x(0) = 230$ and $q = 60$. Using the resulting $x(t)$ for the data y_t in the formulas above

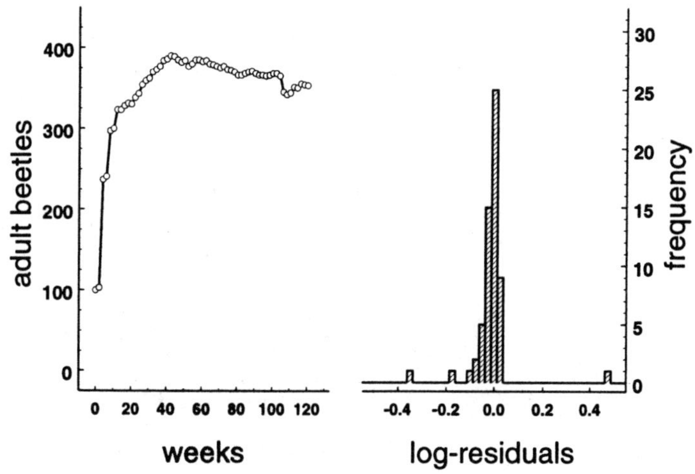

FIG. 1.12. *Adult flour beetle numbers in 20g of flour, shown plotted against time, yield the parameter estimates* $b = 1.631745$, $c = 1.338939 \times 10^{-3}$, $v = 6.658133 \times 10^{-3}$ *for the stochastic Ricker model* (1.52). *The log-residuals* (1.53), *computed from these data and parameter estimates, yield the histogram on the right. The log-residuals have mean* 0.000000 *and variance* 6.770980×10^{-3}.

we obtain (to seven significant digits) the numbers $s_1 = 1.390557 \times 10^4$, $s_2 = 3.911992\times10^6$, $s_3 = -9.005503\times10^{-1}$, $s_4 = -7.101169\times10^3$, $s_5 = 6.893810\times10^1$ and $b = e^{2.3025851} = 1.000000 \times 10^1$, $c = 1.000000 \times 10^{-2}$, $v = 0.000000$. We find that the maximum likelihood parameter estimates accurately recover the correct parameters (seven significant digits).

More interestingly we can apply the parameter estimation formulas above using population census data and see how well the stochastic Ricker model (1.52) accounts for that data. In Fig. 1.12 we show a time series plot of adult flour beetle numbers, one of several replicates in a laboratory experiment in which each replicate was grown in a separate bottle containing 20g of flour. The adult beetles were counted at two week intervals for 120 weeks. Thus, $q = 60$ and there are 61 data points starting with 100 adults at time $t = 0$. Using this data in the formulas above, we obtain the maximum likelihood parameter estimates for b, c, and v given in Fig. 1.12 for the stochastic Ricker model (1.52). How well does this parameterized model do in accounting for the data? The parameterized Ricker model produces, from each data point, a one-step prediction for the next data point. The stochastic model, as a hypothesis to be tested, asserts that the differences between these one-step predictions and the actual data points (the "residuals") have certain statistical properties. Statistical tests can be performed to quantify whether the residuals do indeed have these properties (to a certain level of confidence). The stochastic Ricker model (1.52) predicts that on the log scale the residuals will be normally distributed, i.e., that the

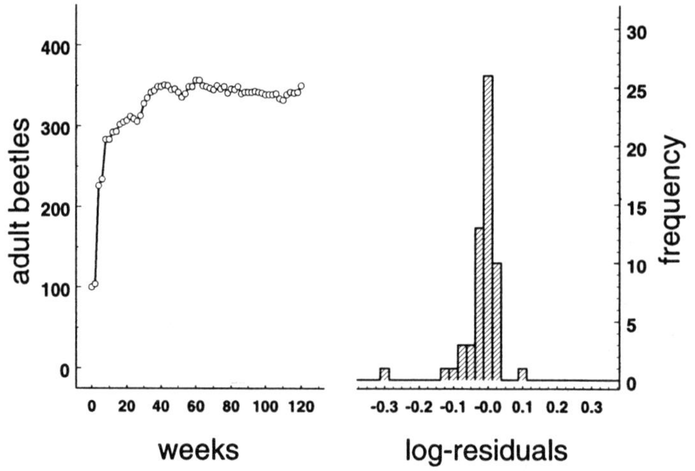

FIG. 1.13. *Beetle numbers from a replicate culture taken from the same experiment that produced the beetle numbers in Fig. 1.12 are shown. The histogram is that of the log-residuals computed from these new data and the parameter estimates calculated from the data in Fig. 1.12. The log-residuals have mean* -3.508827×10^{-2} *and variance* 5.715805×10^{-3}.

log-residuals

$$r_t = \ln\left(\frac{y_t}{y_{t-1}}\right) - \ln b + c y_{t-1} \tag{1.53}$$

will be normally distributed with mean 0 and variance v. For the data and the parameter values in Fig. 1.12 a histogram of these residuals is also plotted. One test of the success of this parameterized model would be to determine whether these residuals are normally distributed with mean 0 and variance $v = 6.658133 \times 10^{-3}$. Normality tests can be performed to see how well the residuals fit this description. Other statistical tests could be carried out as well. (For example, the stochastic Ricker model was built under the assumption that the random variables E_t were uncorrelated in time and hence autocorrelation tests can be performed, on the logarithmic scale, to test this hypothesis.) In this example we will be content with only a visual inspection histogram of log-residuals as shown in Fig. 1.12, which we note has an accurate mean and variance and does have a strong appearance of normality, although there is some skewness to the left.

Suppose we accept that the stochastic Ricker model successfully describes data in Fig. 1.12. A further, and perhaps stronger, test of the model would be to ask that it also describe *other* data gathered under the same experimental conditions. That is, how well does the parameterized model "predict" data that were not used in its parameterization? A plot of data from a replicate culture from the same experiment that yielded the data in Fig. 1.12 appears in Fig. 1.13. Using the parameter estimates obtained from the data in Fig. 1.12, we can calculate log-residuals (1.53) between the Ricker model one-step predictions and the data in Fig. 1.13. The results appear in the histogram in Fig. 1.13. Again

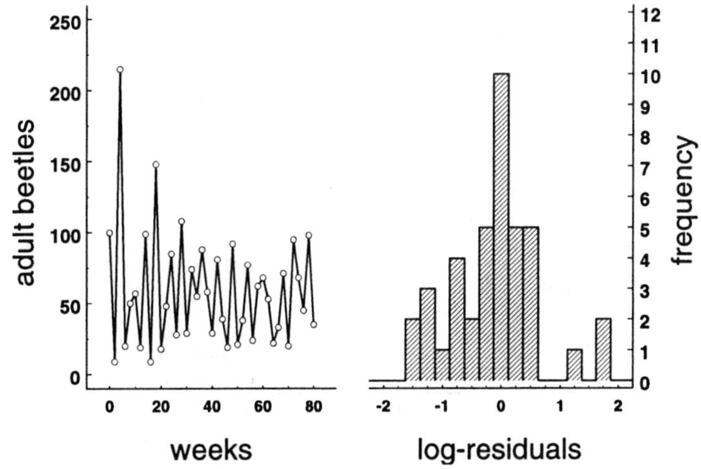

FIG. 1.14. *For the data plotted on the left the maximum likelihood estimates for the stochastic Ricker model* (1.52) *are* $b = 4.815900$, $c = 2.695051 \times 10^{-2}$, *and* $v = 5.362000 \times 10^{-1}$. *The log-residuals, whose histogram is shown on the right, have mean* 0.000026 *and variance* 5.499105×10^{-1}.

we find a normal looking distribution of log-residuals (with some skewness to the left) with accurately predicted mean and variance.

If the "validation" and "prediction" fits in Figs. 1.12 and 1.13 are judged acceptable, then the stochastic Ricker model (1.52), with parameter estimates in Fig. 1.12, is judged adequate to describe the dynamics of the populations in this experiment. Since the parameter estimates in Fig. 1.12 for the deterministic Ricker model yield a stable equilibrium, we would conclude that the populations in this experiment have a "noisy" equilibrium.

In Figs. 1.14 and 1.15 the procedure above is repeated for two more replicate data sets taken from a different experiment. The data are considerably more oscillatory than those in Figs. 1.12 and 1.13. Visually, in this case, the log-residuals are perhaps less convincingly normally distributed. If, however, these validation and prediction fits were judged to be acceptable for these data, then the conclusion would be that the populations in this experiment are adequately described by the stochastic Ricker model (1.52). Since the parameter estimates in Fig. 1.13 for the deterministic Ricker model yield a stable equilibrium, we would conclude that the populations in this experiment also have a "noisy" equilibrium.

A third data set appears in Fig. 1.16 together with the parameter estimates for the stochastic Ricker model and a histogram of the resulting log-residuals. Here the distribution of log-residuals clearly is not normally distributed, and we reject the fit of this model to the data (which asserts that this data is also a very noisy equilibrium). This means either the deterministic Ricker model or the type of noise added to it in (1.52) are inadequate to account for the data in this experiment.

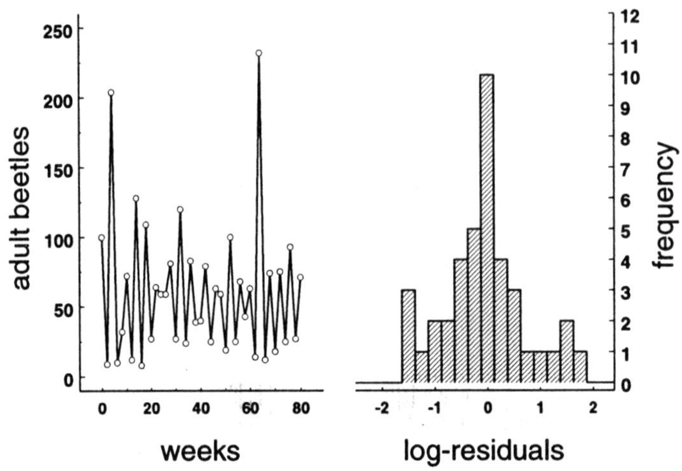

FIG. 1.15. *Beetle numbers from a replicate culture taken from the same experiment that produced the beetle numbers in Fig.* 1.14 *are shown plotted on the left. The histogram is that of log-residuals computed from these data and the parameter estimates calculated from the data in Fig.* 1.14. *The log-residuals have mean* 5.069767×10^{-2} *and variance* 6.178831×10^{-1}.

In the discussion above, the one-dimensional Ricker model was used. The procedures outlined can be extended to the case of multivariate data sets and higher dimensional models [132]. In the multivariate case there can be covariances among the state variables as well as variances. The likelihood function is more complicated in that case, of course, and its maximization must be done numerically with the aid of a computer in order to obtain parameter estimates. It is interesting to note that in the one-dimensional case treated above maximizing the likelihood function is equivalent to minimizing the residuals in the least squares sense; see (1.49). This is not in general the case for the multivariate data (unless all covariances are 0) and parameter estimates calculated by least squares are in general different. Least square estimates relax the distributional assumptions concerning the noise in the model, but should give similar results if the log-normal assumption in the model is valid. In this way, least square estimates provide a cross check on the validity of the stochastic model. However, the statistical theory for multivariate least squares estimates does not seem to have been thoroughly worked out yet.

1.4.2 Bifurcating beetles. The data used in Figs. 1.12–1.16 were taken from laboratory experiments using flour beetles (*Tribolium*). In these experiments, beetles were grown in 20g of flour at a constant incubator temperature and humidity. Every two weeks the flour was shifted and the counted animals returned to a fresh 20g of flour. The data shown in Figs. 1.12–1.16 are numbers of adult beetles taken from three different experiments (in which adult beetle recruitment and mortality were different). While the stochastic Ricker model seemed to describe well some of these data sets, it did less well on others and, in particular, did a very poor job on the data in Fig. 1.16. In order to construct

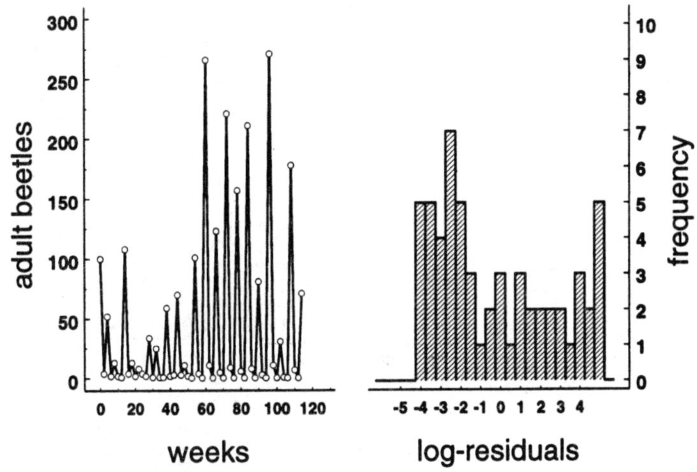

FIG. 1.16. *Beetle numbers from a third experiment are shown plotted on the left. On the right appears a histogram of the log-residuals computed using the parameter estimates* $b = 2.424700$, $c = 2.481400 \times 10^{-2}$, *and* $v = 8.900700$ *obtained from the data. The log-residuals have mean* 4.147428×10^{-2} *and variance* 9.055077.

a better model we should take into account what are considered the important biological mechanisms that effect the dynamics of these beetles. Under the conditions of the experiments (no resource limitation, constant habitat, no other species present, etc.) these mechanisms, it turns out, involve interactions between the individuals of different life cycle stages. The species of flour beetles used in the experiments have a life cycle that includes larval and pupal periods lasting approximately two weeks each (under the conditions of the experiment). Specifically, adult and larvae are cannibalistic, both consuming eggs (affecting larval recruitment). Another significant interaction is adult cannibalism of pupae. Therefore a structured population model is called for that includes at least larval, pupal, and adult stages. (Larvae and pupae were, in fact, also counted in the experiment.)

Since the census period is two weeks, we choose the unit of time to be two weeks and let $L(t)$, $P(t)$, and $A(t)$ denote the numbers of larvae, pupae, and adults at time t, respectively. Given that the larval and pupal periods are both approximately two weeks in duration, a Leslie matrix

$$\begin{bmatrix} 0 & 0 & b \\ 1-\mu_l & 0 & 0 \\ 0 & 1-\mu_p & 1-\mu_a \end{bmatrix}$$

could serve as a projection matrix in the absence of any nonlinear (density) effects. Here b is the larval recruitment rate per adult per unit time (two weeks) and the μ's denote the death rates for the three life-cycle stages. It turns out that for this experiment pupal mortality (in the absence of adult cannibalism) is negligibly small so we set $\mu_p = 0$. Cannibalistic acts are random occurrences as the organisms move through the container of flour. Therefore we model the effects of cannibalism by means of exponential discounting terms of Ricker type (see section 1.1.3). This results in a nonlinear projection matrix of the form

$$P = \begin{bmatrix} 0 & 0 & b\exp\left(-c_{el}L - c_{ea}A\right) \\ 1-\mu_l & 0 & 0 \\ 0 & \exp\left(-c_{ea}A\right) & 1-\mu_a \end{bmatrix}$$

with

$$b > 0, \quad 0 < \mu_l,\, \mu_a < 1, \quad c_{ea},\, c_{el},\, c_{pa} \geq 0.$$

This is the projection matrix for the LPA model equations (1.43), i.e.,

$$\begin{aligned} L(t+1) &= bA(t)\exp\left(-c_{el}L(t) - c_{ea}A(t)\right), \\ P(t+1) &= \left(1-\mu_l\right)L(t), \\ A(t+1) &= P(t)\exp\left(-c_{pa}A(t)\right) + \left(1-\mu_a\right)A(t). \end{aligned} \tag{1.54}$$

The coefficients c_{el}, c_{ea}, and c_{pa} are referred to as the "cannibalism coefficients."

The goal is first to parameterize and validate this deterministic model, using a stochastic version of the model, and then to carry out laboratory experiments based on the dynamics predicted by this model if one or more of its parameters is changed (i.e., manipulated by the experimenter). In particular, it is attempted to document by means of the experiments any bifurcations in the asymptotic dynamics predicted by this deterministic model. The data from the experiments can be analyzed by means of a "prediction fit" (i.e., utilizing the a priori parameter estimates used to design the experiment, not parameter estimates obtained from the resulting experimental data) or, since the experiments include replicate cultures, the data can be divided and different sets used for validation and prediction fits of the model (as in the example of section 1.4.1). We briefly describe the results of such experiments as reported in [132], [72], [73], [116], [133].

To carry out the program described above it is necessary to understand as much as possible about the asymptotic dynamics of the LPA model (1.54). In section 1.3 it was shown that for $n < 1$ the origin is globally attracting in R_+^3 and for $n > 1$ the model system is uniformly persistent and there exists a unique positive equilibrium which is (locally) stable for n sufficiently close to 1. (For a global stability result for the positive equilibrium see [271].) It was pointed out in section 1.3 that the positive equilibria of (1.54) are not stable for all parameter values, however. While there are no general results available that locate stability boundaries for the model parameters in (1.54), numerical simulations, such as those in section 1.3, show that destabilization and bifurcation to nonequilibrium attractors can occur as parameters are varied.

One special case has been thoroughly studied analytically, however. In [132] the destabilization boundaries are analytically found for the case when larval cannibalism of eggs is ignored $c_{el} = 0$. It is shown that positive equilibria can destabilize as a parameter varies either because an eigenvalue of the Jacobian passes through -1 or because a complex conjugate pair of eigenvalues passes out of the complex unit circle. In the first case, a familiar period doubling bifurcation to a 2-cycle occurs while in the second case, a bifurcation to an invariant loop in R^3_+ (and quasi-periodic oscillations) occurs. Which type of destabilization and bifurcation occurs depends on the values of the other model parameters. Extensive numerical studies show this is also the case when $c_{el} \neq 0$. Moreover, these studies show that further destabilizations and bifurcations often occur and chaotic attractors can arise [72], [73], [133].

The richness of dynamics associated with the LPA model (1.54) and the ready availability of flour beetles for laboratory experiments present an opportunity to coordinate theoretical modeling and predictions with experimental data. Moreover, this system offers an opportunity to test the descriptive and predictive capabilities of simple mathematical models (such as (1.54)) for the dynamics of a real biological population and thereby to study the role of nonlinear theory in population dynamics. To connect the LPA model (1.54) to time series data in the manner described in section 1.4.1 a stochastic version of the model is needed.

In [132] the stochastic model

$$\begin{aligned} L(t+1) &= bA(t)\exp\left(-c_{el}L(t) - c_{ea}A(t)\right)\exp\left(E_1(t)\right), \\ P(t+1) &= \left(1-\mu_l\right)L(t)\exp\left(E_2(t)\right), \\ A(t+1) &= \left[P(t)\exp\left(-c_{pa}A(t)\right) + \left(1-\mu_a\right)A(t)\right]\exp\left(E_3(t)\right) \end{aligned}$$

was parameterized and validated using a historical data set. Here the $E_i(t)$ are normal random variables (uncorrelated through time) with mean 0 and variances and covariances to be estimated as model parameters from time series data. Based on that preliminary study and its parameter estimates, an experiment was devised in which the adult death rate μ_a was manipulated in such a way that the dynamics were predicted by the deterministic LPA model (1.54) to undergo a specific sequence of bifurcations. The experiment, which included controls and replicates, lasted 36 weeks. The data from half of the replicates were used for a parameterization (using both maximum likelihood and conditioned least squares methods) and model fit analysis, and the data from remaining replicates were used for a model prediction analysis.

The results are reported in [72] and [133], where details of the laboratory experiments and the parameterization and validation of the model can be found. The maximum likelihood parameter estimates (with 95% confidence intervals) for one of the two different genetic strains of *Tribolium castaneum* (Herbst) used in this experiment appear in Table 1.1. (An estimate for μ_a was obtained from a direct count of dead individuals in the control cultures and found to be 4.210×10^{-3} with a 95% confidence interval of $\left(3.2 \times 10^{-3}, 5.2 \times 10^{-3}\right)$.)

The matrix of variances and covariances of the E_i was estimated for the manipulated treatments to be

$$\begin{bmatrix} 8.280 \times 10^{-1} & -8.448 \times 10^{-2} & -2.971 \times 10^{-3} \\ -8.448 \times 10^{-2} & 1.794 \times 10^{-1} & -2.020 \times 10^{-2} \\ -2.971 \times 10^{-3} & -2.020 \times 10^{-2} & 2.057 \times 10^{-2} \end{bmatrix}.$$

TABLE 1.1

Parameter	ML estimate	Confidence interval
b	7.876	$(5.8, 10.7)$
μ_l	1.613×10^{-1}	$\left(1.0 \times 10^{-1}, 2.2 \times 10^{-1}\right)$
c_{ea}	1.114×10^{-2}	$\left(1.0 \times 10^{-2}, 1.2 \times 10^{-2}\right)$
c_{el}	1.385×10^{-2}	$\left(1.2 \times 10^{-2}, 1.5 \times 10^{-2}\right)$
c_{pa}	4.348×10^{-3}	$\left(3.6 \times 10^{-3}, 4.8 \times 10^{-3}\right)$

With these parameter values the LPA model equations (1.54) predict a 2-cycle bifurcation as μ_a is increased through a critical value of approximately 0.1 as shown in the bifurcation diagram of Fig. 1.17. The data obtained from experiments performed at adult death rates $\mu_a = 0.04$, 0.27, and 0.50 are shown in Fig. 1.18 plotted in three-dimensional phase space together with the model predicted attractors [133]. Clearly these data support the predicted 2-cycle bifurcation.

Experiments of the type described above were repeated, for a longer length of time, in which both adult mortality (held at $\mu_a = 0.96$) and adult recruitment (i.e., c_{pa}) were manipulated. The model predicted bifurcations as c_{pa} is increased (with μ_a held at 0.96 and other parameter values taken from Table 1.1) are seen in Fig. 1.19 to be considerably more complicated than those in Fig. 1.17. A stable equilibrium exists at $c_{pa} = 0$, but a discrete Hopf (invariant loop) bifurcation occurs at a very small positive value of c_{pa}. For example, an invariant loop occurs at $c_{pa} = 0.05$. As c_{pa} is further increased invariant loops are interspersed with "windows" where periodic attractors of high period occur. Ultimately, chaotic attractors appear. For example at $c_{pa} = 0.35$ the attractor has Liapunov exponent 0.1146 and box dimension approximately 1.25. For large values of c_{pa} there is a distinctive 3-cycle. Eighty weeks of data (with transients left out) from four of the eight treatments reported in [73] are plotted in Fig. 1.20 together with the corresponding model predicted attractor. In this case, none of the data is used for parameter estimation (so that the results deal with the *prediction* capabilities of the previously fitted and validated model)! The model attractors are computed from the parameter estimates in Table 1.1 obtained from the earlier experiment. Again one sees a convincingly close match between the experimental data and the deterministic model predicted sequence of bifurcations. These results are the first convincing evidence of the possibility of chaotic dynamics in a biological population. See [62], [185], [255], [360].

1.4.3 Periodic habitats. Autonomous models assume a constant environment. If environmental fluctuations cause a model's parameters to vary in time, then the model becomes nonautonomous. For example, if fertility and mortality rates are seasonal, then it might be reasonable to consider a periodically forced model in which fertility and mortality coefficients are periodic functions of time. Since regular environmental fluctuations (seasonal, daily, etc.) are

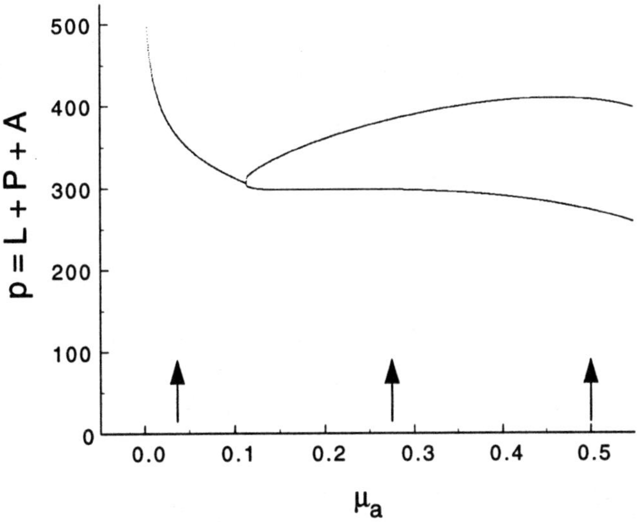

FIG. 1.17. *A bifurcation diagram, plotting total population size p against the adult death rate μ_a, for the flour beetle model* (1.54) *using parameter values from Table* 1.1 *shows a 2-cycle bifurcation. Arrows indicate where laboratory experiments were performed:* $\mu_a = 0.04$, 0.27, 0.50.

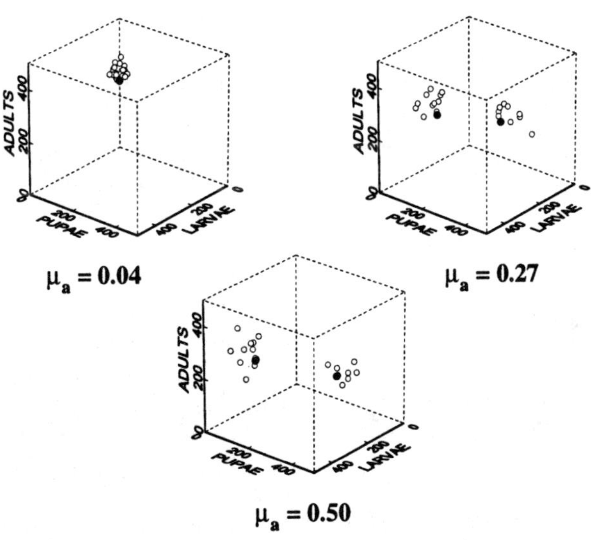

FIG. 1.18. *Open circles show data (with transients removed) taken from laboratory experiments using flour beetles with adult death rates μ_a manipulated to the indicated values. Solid circles show the model predicted attractor.*

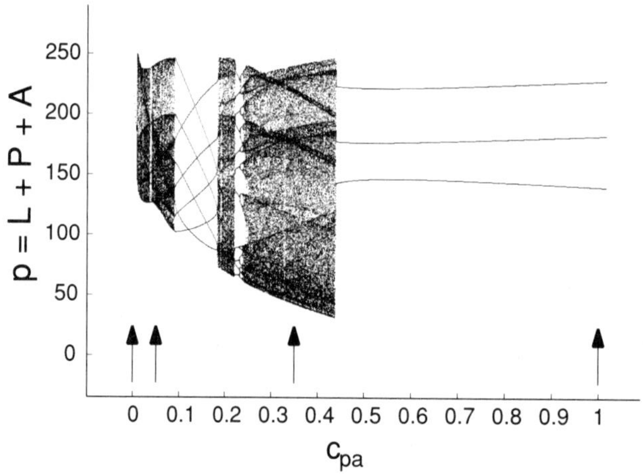

FIG. 1.19. *The bifurcation diagram of the flour beetle model* (1.54) *using parameter values from Table* 1.1 *with* $\mu_a = 0.96$. *Total population size is plotted against* c_{pa}. *Arrows indicate where laboratory experiments were performed.*

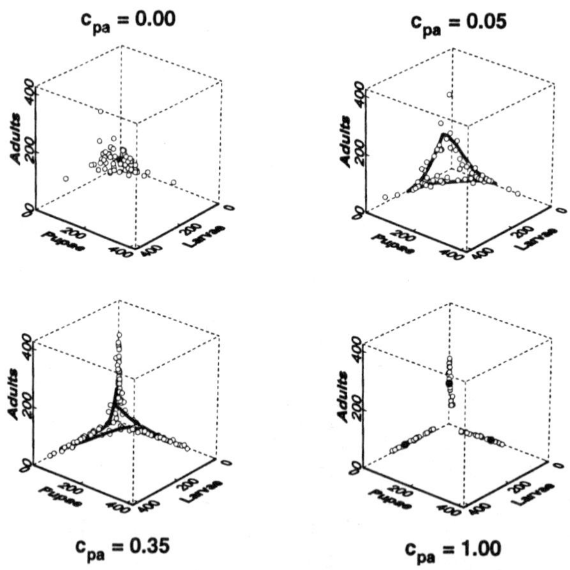

FIG. 1.20. *Open circles show data (with transients removed) taken from experiments in which* c_{pa} *was manipulated. Also plotted are the model* (1.54) *predicted attractors, which are: an equilibrium for* $c_{pa} = 0.00$, *an invariant loop for* $c_{pa} = 0.05$, *a chaotic attractor for* $c_{pa} = 0.35$, *and a* 3-*cycle for* $c_{pa} = 1.0$.

certainly important for the dynamics of many species, one might expect to find more periodically forced models in the literature. While there is a small body of literature on periodically forced models in population dynamics and ecology, the vast majority of models are autonomous. (This is no doubt due to the fact that autonomous models are more tractable than periodically forced models.) An interesting general question is, how many of the fundamental tenets of theoretical ecology remain valid in a periodically (or otherwise) fluctuating environment? See the references cited in [232] for literature on periodically forced models.

An opportunity to study a population in a periodically fluctuating environment is provided by the results of some laboratory experiments with flour beetles performed by Jillson [252]. In these experiments the same experimental protocol described above was used, except that the beetles were not always returned to 20g of flour. Instead the flour amount was changed at each census time (2 weeks) in a periodic manner so that the *average* amount was 20g. One experiment, for example, involved a flour amount schedule 32g-8g-32g-8g$\cdots$ with period 2, i.e., a period of four weeks. Other experiments involved longer periods. All were carried out with replicates and constant habitat controls.

Given the success of the autonomous model (1.54) in a habitat consisting of a constant amount of flour, it seems reasonable to hope that an appropriate, periodically forced, modification of this model might accurately describe the data in Jillson's experiments, explain any interesting and unusual aspects in this data, and even predict dynamical phenomena that might be successfully observed in new flour beetle experiments. In fact, these expectations have been remarkably well fulfilled.

The following periodically forced version model (1.54) was proposed and studied in [232], [74] as a model for flour beetle dynamics in a fluctuating amount of flour with period 2:

$$\begin{aligned}
L(t+1) &= bA(t)\exp\left(-\frac{c_{el}}{1+\alpha(-1)^t}L(t)-\frac{c_{ea}}{1+\alpha(-1)^t}A(t)\right),\\
P(t+1) &= (1-\mu_l)\,L(t), \qquad\qquad (1.55)\\
A(t+1) &= P(t)\exp\left(-\frac{c_{pa}}{1+\alpha(-1)^t}A(t)\right)+(1-\mu_a)\,A(t).
\end{aligned}$$

This model is built upon the assumption that the cannibalism coefficients (i.e., the probability of a cannibalistic event) are inversely proportional to the amount of flour. This assumption has been validated by laboratory experiments [74]. For example we write $c_{el}=\kappa_{el}/V$, where V is the volume occupied by 20g of flour under laboratory conditions. If $\overline{V}$ is the average flour volume, then in the autonomous model $c_{el}=\kappa_{el}/\overline{V}$. If, however, the flour volume fluctuates with an average $\overline{V}$, a period of 2 and a relative amplitude of α $(0\le\alpha<1)$, then $V=\overline{V}\,(1+\alpha(-1)^t)$ and the larval/egg cannibalism coefficient becomes $\kappa_{el}/\overline{V}\,(1+\alpha(-1)^t)=c_{el}/\,(1+\alpha(-1)^t)$. A similar derivation, when applied to the other cannibalism coefficients, results in the *periodic LPA model* (1.55). The autonomous model (1.54) corresponds to $\alpha=0$.

An interesting consequence of periodically fluctuating flour amounts was reported in [252]. For the experiment in which the period was 2, the total population size was significantly *larger* than that found in the constant habitat control with the same average amount of flour (20g). This surprising result is contrary to an often held theoretical tenet asserting deleterious effects of habitat fluctuations. This tenet is based upon studies of periodically forced, unstructured models, such as the classical logistic equation [309], [331]. (See, however, [92], [364].) Can the periodic LPA model explain this unexpected result?

The periodic LPA model (1.55) is analytically studied in [232] by means of bifurcation and perturbation techniques. (Also see [107] and [227].) A global continuum of positive 2-cycles bifurcates from the extinction equilibrium at the critical value $b_{cr} = \mu_a/(1-\mu_l)$ of the parameter b. This is the same bifurcation point as that for the continuum of positive equilibria in the autonomous LPA model (1.54). The reason for this is that the Jacobians of these two models evaluated at the extinction equilibrium are identical. The continuum of positive 2-cycles bifurcates to the right and is stable. These bifurcation results can be proved using the same general techniques used for the autonomous model. Liapunov/Schmidt expansions can be utilized to compare the (average) total population sizes for the periodic and the autonomous LPA models near the bifurcation point. Let $(\overline{L}_\alpha, \overline{P}_\alpha, \overline{A}_\alpha)$ denote the averages of a 2-cycle solution $(L_\alpha(t), P_\alpha(t), A_\alpha(t))$ of the periodic LPA model (1.55) with amplitude $\alpha < 1$, i.e., $\overline{L}_\alpha = \sum_{t=0}^{1} L_\alpha(t)/2$, etc. For a small parameter $\varepsilon_\alpha > 0$ the Liapunov/Schmidt expansion near the bifurcation point yields [232]

$$\begin{aligned} L_\alpha(t) &= \varepsilon_\alpha b_{cr} + O\left(\varepsilon_\alpha^2\right)(t), \\ P_\alpha(t) &= \varepsilon_\alpha \mu_a + O\left(\varepsilon_\alpha^2\right)(t), \\ A_\alpha(t) &= \varepsilon_\alpha + O\left(\varepsilon_\alpha^2\right)(t), \end{aligned}$$

and

$$b = b_{cr} + \varepsilon_\alpha b_{cr}\left(\frac{c_{ea} + b_{cr}c_{el} + c_{pa}}{1-\alpha^2}\right) + O\left(\varepsilon_\alpha^2\right).$$

We wish to compare the averages of $L_\alpha(t)$, $P_\alpha(t)$, and $A_\alpha(t)$ for $\alpha = 0$ to those for $\alpha > 0$ for the same parameter values. To obtain the same value of b to first order, we take $\varepsilon_\alpha = \varepsilon_0\left(1-\alpha^2\right)$ and find that

$$\begin{aligned} \overline{L}_0 - \overline{L}_\alpha &= \varepsilon_\alpha \alpha^2 b_{cr} + +O\left(\varepsilon_\alpha^2\right), \\ \overline{P}_0 - \overline{P}_\alpha &= \varepsilon_\alpha \alpha^2 \mu_a + +O\left(\varepsilon_\alpha^2\right), \\ \overline{A}_0 - \overline{A}_\alpha &= \varepsilon_\alpha \alpha^2 + +O\left(\varepsilon_\alpha^2\right). \end{aligned}$$

Thus, for small values of b near b_{cr} we have $\overline{L}_0 > \overline{L}_\alpha$, $\overline{P}_0 > \overline{P}_\alpha$, and $\overline{A}_0 > \overline{A}_\alpha$. This *negative effect* of habitat fluctuation, in which the average total population size $\overline{p}_\alpha \doteq \overline{L}_\alpha + \overline{P}_\alpha + \overline{A}_\alpha$ decreases in a periodic habitat, is the *opposite* of Jillson's findings. It is shown analytically in [232] that a *positive effect* $\overline{p}_\alpha > \overline{p}_0$ can occur for the periodic LPA model (1.55) for sufficiently large values of $b > b_{cr}$, at least for small relative amplitudes α. This is done by calculating

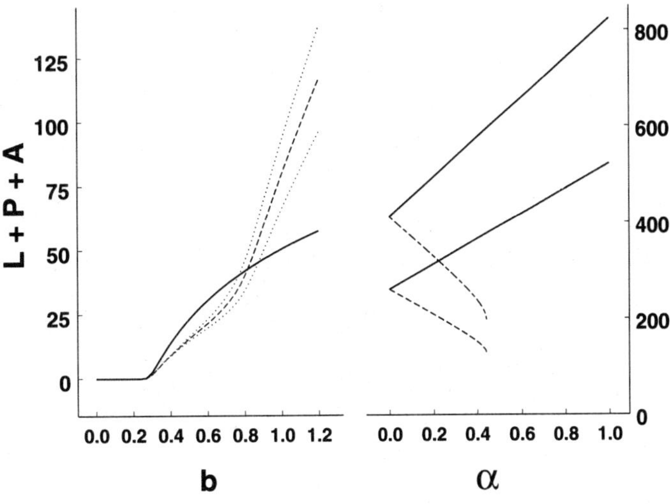

FIG. 1.21. *The left graph shows the stable equilibria of the autonomous LPA model (solid line) and the stable* 2*-cycles of the periodic LPA model (dotted lines) plotted against* b*. The remaining parameters are given by* (1.56)*, and the relative amplitude is* $\alpha = 0.6$ *(as in Jillson's experiment). The averages of the* 2*-cycles (dashed line) exceed the stable equilibria for* b *greater than approximately* $0.8 > 0.293 = b_{cr}$*. In the right graph the stable* 2*-cycles of the periodic LPA model with parameter estimates* (1.56) *are plotted against the relative amplitude* α*. For* $\alpha > 0$ *less than approximately* 0.45 *there exist two stable* 2*-cycles.*

lower order terms in an α expansion of the 2-cycle solution near an equilibrium for $\alpha = 0$. This result shows that the periodic LPA model has at least the theoretical capability of predicting positive effects of habitat fluctuations. See Fig. 1.21. However, parameter estimates computed from the control treatments in Jillson's experiment yield the values

$$(1.56) \qquad \begin{aligned} b &= 4.445, \quad \mu_l = 0.4794, \quad \mu_a = 0.1524, \\ c_{el} &= 0.0584, \quad c_{ea} = 0.0058, \quad c_{pa} = 0.0105. \end{aligned}$$

For these parameter values the autonomous LPA model (1.54) has an unstable equilibrium and a stable 2-cycle. Thus, the 2-cycles for small amplitudes α obtained in [232] are unstable and cannot explain Jillson's result.

In [74] the periodic LPA model (1.55) is used to explain the positive effect observed in Jillson's experiment. When, as for the parameter estimates (1.56), the autonomous LPA model has a stable 2-cycle, perturbation methods show that stable 2-cycles for the periodic LPA model exist for (at least small) relative amplitudes $\alpha > 0$. In fact, there are two stable 2-cycles for small α corresponding to the two phase shifted 2-cycles of the autonomous LPA model. For the periodic LPA model these two stable 2-cycles are not phase shifts of one another and they have different amplitudes and, importantly, different averages.

In Fig. 1.21 the stable 2-cycles are plotted as a function of α for the periodic LPA model with the parameter estimates (1.56). It is seen from these graphs that one 2-cycle has an average total population size greater than that of the 2-cycle in the constant habitat ($\alpha = 0$) while the other has a smaller average total population size. Thus, the periodic LPA model, parameterized to the Jillson experiment, predicts two stable 2-cycles, one implying a positive effect and the other implying a negative effect. However, as can be seen in Fig. 1.21, as α is increased the stable 2-cycle with the smaller average abruptly disappears, leaving only the stable 2-cycle with the larger average. (It is conjectured that a saddle-node bifurcation causes the disappearance of the smaller average 2-cycle as it "collides" with the unstable 2-cycle that emanates from the unstable equilibrium.) For $\alpha = 0.6$, as in Jillson's experiment, the periodic LPA model predicts only the stable 2-cycle with the larger average, and in this way the model explains the positive effect observed by Jillson.

An interesting result of the above analysis of the periodic LPA model is the prediction of two stable attractors for small relative amplitudes $\alpha > 0$. This is unexpected result has been recently confirmed by laboratory experiments [75].

1.5 Multispecies interactions

To this point we have considered structured models for the dynamics of only a single population or species. Matrix models for the interaction of several structured species can be readily constructed by assuming that some entries in the fertility and/or transition matrices of each species depend upon (at least some) class densities of other species. In multispecies models the structuring variables do not have to be of the same kind nor do the numbers of categories used have to be the same for all species. Coupled matrix equations of this kind can be used to model interactions, such as predation and competition, that are specific to structuring classes. For example, only certain classes of individuals might be vulnerable to predation and among those class predation rates might vary considerably (e.g., younger and older individuals are often much more susceptible to predation). Similarly, only certain classes of individuals (e.g., juveniles) might experience competitive pressure from other species for resources. Individuals from two species may compete when small in size (young in age), but not when they are larger (older) [435]. Indeed, structured models can account for interactions that defy the classical typecasting for multispecies interactions. In [435] many examples are given in which species exhibit mixed competition/predation interactions in which individuals of a species, when small (young), compete with those of another species on whom they eventually will prey when larger (older). Such class specific interactions can profoundly affect the population level dynamics of a community of species.

In this section coupled matrix equations will be considered from a bifurcation theory point of view. The approach will be such that the methods and general theorems used for the single species models are applicable to multispecies models.

Consider a community of one or more "species" whose state variables are contained in a column matrix $y = y(t) \in R_+^{m_2}$. A "species" here can be a biological species or a nonbiological quantity such as a food resource. Assume that this community has an equilibrium $y_e > 0$. To this "resident" community we add a (possibly) structured species with state variables $x = x(t) \in R_+^{m_1}$, $m_1 \geq 1$, whose dynamics are described by a matrix model of the type considered in this chapter with a projection matrix P. The species x and the species in the community y are assumed to interact so that $P = P(x, y)$. Thus, we are led to consider a system of the form

$$\begin{aligned} x(t+1) &= P\left(x(t), y(t)\right) x(t), \\ y(t+1) &= g\left(x(t), y(t)\right). \end{aligned} \tag{1.57}$$

Here the community dynamics in the absence of the species x are governed by the equation $y(t+1) = g(0, y(t))$ and $g(0, y_e) = y_e$. If the resident community consists of one or more structured populations, then the dynamics g are given by a projection matrix so that in place of (1.57) we have

$$\begin{aligned} x(t+1) &= P\left(x(t), y(t)\right) x(t), \\ y(t+1) &= Q\left(x(t), y(t)\right) y(t). \end{aligned}$$

This in turn could be written as a single matrix equation by combining the vectors x and y and creating a block diagonal projection matrix with P and Q along the diagonal.

1.5.1 Some equilibrium theory. The system (1.57) has the "trivial" equilibrium $y = y_e > 0$, $x = 0$. We will refer to this as the "extinction" equilibrium, meaning by this that x is absent in this equilibrium state. Our approach to the existence of positive equilibria, in which x is therefore not extinct, is to establish their bifurcation from the extinction equilibrium as a model parameter, related to the "inherent" dynamics of species x, is varied. This will be done in a manner analogous to that used for the single species case in this chapter. Although a global result analogous to Theorem 1.2.4 by an extension of the proof of Theorem B.1.1 (see Appendix B.1) is possible, we will present only simpler local bifurcation results.

$P(0, y_e)$ is called the *inherent projection matrix for x at y_e*. As in section 1.2 we assume that a bifurcation parameter λ of interest appears linearly in this matrix. Thus, we consider the system

$$\begin{aligned} x(t+1) &= P\left(\lambda, x(t), y(t)\right) x(t), \\ y(t+1) &= g\left(x(t), y(t)\right), \end{aligned} \tag{1.58}$$

where for some integer $k \in I[1, +\infty)$

$$\begin{aligned} P(\lambda, y, x) &\doteq [p_{ij}(\lambda, x, y)], \\ p_{ij} &\in C^{k+1}\left(R^1 \times \Omega, R_+^1\right), \ i, j \in I[1, m_1], \\ g &\in C^{k+1}\left(\Omega, R_+^{m_2}\right), \end{aligned}$$

where Ω is an open set containing R_+^m, $m = m_1 + m_2$, and

$$\begin{aligned} g(0, y_e) &= y_e > 0, \\ P(\lambda, 0, y_e) &= A + \lambda B. \end{aligned}$$

The equilibrium equations associated with (1.58) are

$$\text{(1.59)} \qquad \begin{aligned} &\text{(a)} \quad x = P(\lambda, x, y)\, x, \\ &\text{(b)} \quad y = g(x, y). \end{aligned}$$

By assumption $y = y_e$ is an equilibrium of

$$\text{(1.60)} \qquad y(t+1) = g(0, y(t)).$$

We say the equilibrium y_e is *nondegenerate* if 1 is not an eigenvalue of the Jacobian $J_y g(0, y_e)$ of g with respect to y evaluated at $x = 0$, $y = y_e > 0$. Note that if all eigenvalues of $J_y g$ satisfy $|\zeta| < 1$, and hence y_e is (locally asymptotically) stable in the absence of x, then y_e is nondegenerate.

If y_e is nondegenerate, then by the implicit function theorem equation (1.59b) can be solved for $y = y(x)$. Specifically, there exists such a solution $y \in C^{k+1}\left(N(0), R^{m_2}\right)$, $y(0) = 0$, where $N(0)$ is an open neighborhood of 0 in R^{m_1}. For $x \in N(0)$ equation (1.59a) becomes $x = P\left(\lambda, x, y(x)\right)x$ or $x = (A + \lambda B)\, x + r\left(\lambda, x\right)$, where $r(\lambda, x) = O(|x|^2)$ near $x = 0$. If $I - A$ is invertible, then this equation can be written in the form $x = \lambda L x + h(\lambda, x)$, where $L \doteq (I - A)^{-1} B$ and $h(\lambda, x) \doteq (I - A)^{-1} r(\lambda, x)$ to which Theorem B.2.1 (see Appendix B.2) can be applied. The result is that Theorems 1.2.5 and 1.2.6 are valid concerning the bifurcation of solutions x from 0 at a critical value λ_0 of $A + \lambda B$. Coupling these results with the solution $y = y(x)$ we get the following local bifurcation result for the equilibria of the multispecies system (1.58).

THEOREM 1.5.1. *Suppose that λ_0 is a critical value of $A + \lambda B$, $I - A$ is invertible, and $w^\tau B v \neq 0$. Suppose that $y_e > 0$ is a nondegenerate equilibrium of* (1.60). *Then a branch of positive equilibria x and y of the multispecies matrix model* (1.58) *bifurcates from the extinction equilibrium $x = 0$ and $y = y_e > 0$. This branch has the form*

$$\lambda = \lambda_0 + \eta(\varepsilon), \quad x = \varepsilon v + z(\varepsilon), \quad y = y_e + \left(I - J_y g(0, y_e)\right)^{-1} J_x g(0, y_e) v \varepsilon + w(\varepsilon),$$

where for some $\varepsilon_0 > 0$

$$\begin{aligned} &\eta \in C^k\left((-\varepsilon_0, \varepsilon_0), R^1\right), \quad \eta(0) = 0, \\ &z \in C^k\left((-\varepsilon_0, \varepsilon_0), V^\perp\right), \quad z(0) = z'(0) = 0, \\ &w \in C^k\left((-\varepsilon_0, \varepsilon_0), R^{m_2}\right), \quad w(0) = w'(0) = 0. \end{aligned}$$

Moreover, if $k \geq 2$, then

$$\lambda_1 = \eta'(0) = -\frac{w^\tau \left[d_{ij}\right] v}{w^\tau B v},$$

$$d_{ij} = \left(\nabla_x p_{ij}(\lambda_0, 0, y_e)\right)^\tau v + \left(\nabla_y p_{ij}(\lambda_0, 0, y_e)\right)^\tau \left(I - J_y g(0, y_e)\right)^{-1} J_x g(0, y_e) v.$$

The formula for $\eta'(0)$ can be obtained by substituting the expansions for λ, x, and y in Theorem 1.5.1 into equation $x = P(\lambda, x, y(x))\,x$, differentiating the twice with respect to ε and setting $\varepsilon = 0$ in the result. This yields the equation

$$(I - (A + \lambda_0 B))\, z''(0) = 2\,(B\eta'(0) + [d_{ij}])\, v$$

for $z''(0)$. Since the null space of $I - (A + \lambda_0 B)$ is spanned by v, it is necessary for the solution of this equation that the right-hand side be orthogonal to w.

We turn now to the stability of the extinction equilibrium and of the positive equilibria in Theorem 1.5.1. The Jacobian $J(\lambda, x, y)$ of (1.58) evaluated at the extinction equilibrium $x = 0$, $y = y_e$ is the block diagonal matrix

$$J(0, y_e) = \begin{bmatrix} A + \lambda B & 0 \\ J_x g(0, y_e) & J_y g(0, y_e) \end{bmatrix}.$$

The eigenvalues of $J(0, y_e)$ are those of the individual blocks $A + \lambda B$ and $J_y g(0, y_e)$. Thus, if $J_y g(0, y_e)$ has an eigenvalue with $|\zeta| > 1$, then the extinction equilibrium is unstable for all λ.

Let us consider the problem of a species x trying to "invade" a community which is at stable equilibrium. By "invade" we mean that there is a positive stable equilibrium. Since we are considering only local bifurcations of positive equilibria from the extinction state (hence small amplitude x equilibrium states) and only local stability, the ecological problem is that of population x attempting to invade a community at low population numbers. The community y is supposed stable in the absence of x, so it is assumed that all eigenvalues of $J_y g(0, y_e)$ satisfy $|\zeta| < 1$. The stability of the extinction equilibrium is therefore determined by the magnitude of the eigenvalues of block $A + \lambda B$. Theorem 1.2.3 applies and tells us that the extinction equilibrium loses stability as either λ increases ($w^\tau Bv > 0$) or decreases ($w^\tau Bv < 0$) through the critical value λ_0.

We turn now to the local stability of the positive equilibria given in Theorem 1.5.1. Based on the results for single species matrix models we expect that stability is related to the direction of bifurcation and hence to the sign of λ_1. An analog of Lemma 1.2.2 can be proved by a straightforward, but tedious, calculation. Thus, if $\zeta = \zeta(\varepsilon)$ is the dominant eigenvalue of the Jacobian of $J(\lambda, x, y)$ evaluated at the positive equilibria in Theorem 1.5.1, then it turns out that

$$\zeta'(0) = -\kappa \frac{w^\tau Bv}{w^\tau v} \lambda_1$$

for a positive constant $\kappa > 0$.

THEOREM 1.5.2. *Suppose, in addition to the assumptions of Theorem* 1.5.1, *that all eigenvalues of $J_y g(0, y_e)$ satisfy $|\zeta| < 1$ (so $y_e > 0$ is stable). The extinction equilibrium $x = 0$, $y = y_e$ of the multispecies matrix model* (1.58) *loses stability as λ increases through λ_0, if $w^\tau Bv > 0$, or as λ decreases through λ_0, if $w^\tau Bv < 0$.*

Suppose further that λ_0 is a strictly dominant critical value, $k \geq 2$ and $\lambda_1 \neq 0$. If $w^\tau Bv > 0$, then the bifurcation described in Theorem 1.5.1 *is stable if it is to* *the right* ($\lambda_1 > 0$) *and unstable if it is to the left* ($\lambda_1 < 0$). *If* $w^\tau Bv < 0$, *then the bifurcation is stable if it is to the left and unstable if it is to the right.*

One choice of the parameter λ is the *inherent net reproductive number* n *of species* x *at the extinction equilibrium* $(0, y_e)$. Write

$$P(n, x, y) = T(x, y) + n\Phi(x, y),$$

where $T(x, y)$ is the transition matrix and $n\Phi(x, y)$ is the fertility matrix for species x. Here Φ is normalized so that n is the inherent net reproductive number for the species x when y is at the equilibrium y_e; i.e., 1 is the dominant eigenvalue of $T(0, y_e) + \Phi(0, y_e)$. In this case $B = \Phi(0, y_e) \geq 0$ ($\neq 0$) and $w^\tau Bv > 0$ so that the extinction equilibrium loses stability as n is increased through 1 and the bifurcation is stable if and only if it is to the right, i.e., $w^\tau [d_{ij}] v < 0$.

As in the case of a single species model, the stability properties in a neighborhood of the bifurcation point may change outside that neighborhood through various types of bifurcations depending on how the eigenvalues of the Jacobian leave the unit circle.

Remark. Theorems 1.5.1 and 1.5.2 remain valid as stated if λ appears in $g = g(\lambda, x, y)$ and if there exists an equilibrium $y_e > 0$ independent of λ (at least for λ near λ_0). In this case y_e nondegenerate means 1 is not an eigenvalue of the Jacobian $J_y g(\lambda_0, 0, y_e)$.

1.5.2 Applications

Interactions of two structured species. Consider two interacting species described by the coupled matrix equations

$$\begin{aligned} x(t+1) &= P_1(p_1(t), p_2(t))x(t), \\ y(t+1) &= P_2(p_1(t), p_2(t))y(t), \end{aligned}$$

where

$$p_1(t) = \sum_{i=1}^{m_1} \omega_i x_i(t), \quad p_2(t) = \sum_{i=1}^{m_2} \chi_i y_i(t)$$

are weighted total population sizes of $x(t)$ and $y(t)$. The projection matrices

$$P_i(p_1, p_2) = T_i(p_1, p_2) + F_i(p_1, p_2)$$

are assumed, in addition to satisfying the conditions in (1.58), to be irreducible and primitive and to have net reproductive numbers $n_i(p_1, p_2)$ at equilibrium (p_1, p_2) (the positive, strictly dominant, simple eigenvalues values of $F_i(p_1, p_2)(I - T_i(p_1, p_2))^{-1}$). As in section 1.2.5, at equilibrium we have $n_i(p_1, p_2) = 1$ for both $i = 1, 2$ (see (1.32)).

Let $P = P_1(p_1, p_2)$, $g = P_2(p_1, p_2)y$ in (1.58). Assume that species y has a stable equilibrium in the absence of species x. That is to say, we assume that $y(t+1) = P_2(0, p_2(t))y(t)$ has a stable positive equilibrium $y_e > 0$ (i.e., the eigenvalues ζ of the Jacobian of the right-hand side evaluated at y_e satisfy $|\zeta| < 1$). Thus, we view y as a "resident" species in stable equilibrium and x as a species trying to "invade" at low density. Denote by $p_2^e > 0$ the weighted total population size of the equilibrium y_e, and write

$$P_1(p_1, p_2) = T_1(p_1, p_2) + n\Phi_1(p_1, p_2),$$

where n is the *inherent* net reproductive number of species x at the *extinction* equilibrium $p_1 = 0$, $p_2 = p_2^e > 0$ and Φ_1 is the normalized fertility matrix (so 1 is the dominant eigenvalue of $T_1(0, p_2^e) + n\Phi_1(0, p_2^e)$). Use $\lambda = n$ in Theorems 1.5.1 and 1.5.2.

The direction of bifurcation at $n = 1$ can be related to the properties of the net reproductive numbers n_i as follows. Assume that both species have density self regulation in the sense that

$$\partial_{p_1} n_1(p_1, p_2) < 0, \quad \partial_{p_2} n_2(p_1, p_2) < 0.$$

Since $n_1(p_1, p_2) = n\nu_1(p_1, p_2)$, where ν_1 is the net reproductive number of $T_1(p_1, p_2) + \Phi_1(p_1, p_2)$, at equilibrium we have

$$\begin{aligned} n\nu_1(p_1, p_2) &= 1, \\ n_2(p_1, p_2) &= 1. \end{aligned} \tag{1.61}$$

For the equilibria in Theorem 1.5.1 the bifurcation will be to the right (and stable) if $\nu_1(p_1, p_2) < \nu_1(0, p_2^e) = 1$ for (p_1, p_2), $p_1 > 0$, close to $(0, p_2^e)$. Using the implicit function theorem we can solve the second equation in (1.61) for $p_2 = p_2(p_1)$, $p_2(0) = p_2^e$ and substitute the answer into the first equation in (1.61) to obtain

$$n\nu_1(p_1, p_2(p_1)) = 1.$$

The bifurcation will be to the right (and stable) if the factor $\nu_1(p_1, p_2(p_1))$ is decreasing at $p_1 = 0$, i.e.,

$$\frac{d}{dp_1}\nu_1(0, p_2^e) = \partial_{p_1}\nu_1(0, p_2^e) + \partial_{p_2}\nu_1(0, p_2^e)p_2'(0) < 0.$$

Since $p_2'(0) = -\partial_{p_1} n_2(0, p_2^e)/\partial_{p_2} n_2(0, p_2^e)$, this inequality is the same as

$$\det M > 0, \quad M \doteq \begin{bmatrix} \partial_{p_1} n_1 & \partial_{p_2} n_1 \\ \partial_{p_1} n_2 & \partial_{p_2} n_2 \end{bmatrix},$$

where the partial derivatives are evaluated at $(p_1, p_2) = (0, p_2^e)$ and $n = 1$. On the other hand, $\det M < 0$ implies that the bifurcation is to the left (and unstable). See Fig. 1.22.

We can define the two most fundamental types of ecological interactions using the net reproductive numbers of the species in the following way:

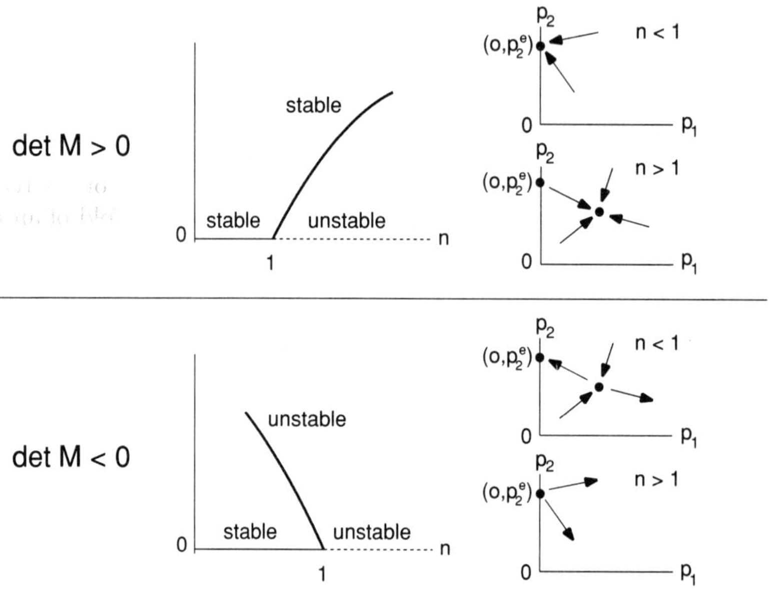

FIG. 1.22. *The bifurcation scenarios for a two species interaction model are schematically represented by bifurcation diagrams on the left and phase portraits on the right. In the bifurcation diagrams the distance from the positive equilibria* (p_1, p_2) *to the extinction equilibrium* $(0, p_2^e)$ *is plotted against the inherent net reproductive number* n *of species* p_1 *at the equilibrium* $(0, p_2^e)$.

$$\text{predator–prey:} \quad \partial_{p_2} n_1(p_1, p_2) > 0, \quad \partial_{p_1} n_2(p_1, p_2) < 0,$$

$$\text{competition:} \quad \partial_{p_2} n_1(p_1, p_2) < 0, \quad \partial_{p_1} n_2(p_1, p_2) < 0.$$

In a predator–prey interaction, the invading species p_1 is the predator and the resident species p_2 is the prey. In this case $\det M > 0$ and the predator can successfully invade (i.e., coexist in stable equilibrium with) the prey at low densities provided the predator's inherent net reproductive number n at the prey equilibrium $(0, p_2^e)$ exceeds 1. If, on the other hand, $n < 1$, then low population densities of the predator will go extinct. See Fig. 1.22.

In a competition interaction $\det M = \partial_{p_1} n_1 \partial_{p_2} n_2 - \partial_{p_2} n_1 \partial_{p_1} n_2$ can be of either sign, depending upon the relative magnitude of the product $\partial_{p_1} n_1 \partial_{p_2} n_2$ (as a measure of the strength of *intra*specific competition in the system) and the product $\partial_{p_2} n_1 \partial_{p_1} n_2$ (as a measure of the strength of the *inter*specific competition). In either case, low population densities of the invading species p_1 will go extinct if its inherent net reproductive number n at the resident competitor's equilibrium $(0, p_2^e)$ is less than 1. If $n > 1$, then species p_1 can successfully "invade" the resident competitor species p_2 at low population densities in the sense that low initial population densities of p_2 will not go extinct. If intraspecific competition outweighs interspecific competition (i.e., $\det M > 0$), then the two species will coexist in stable equilibrium if $n > 1$. If interspecific competition outweighs intraspecific competition (i.e., $\det M < 0$), then there is no stable positive equilibrium near the bifurcation point and the asymptotic outcome when $n > 1$ cannot in general be determined from a local bifurcation analysis. In this case it is often true in competition models that all orbits tend to an axis equilibrium (except perhaps for those on a stable manifold of an unstable equilibrium) so that one species always goes extinct, in keeping with the celebrated competitive exclusion principle in ecology. However, in this case it is also possible for two competing species coexist in a nonequilibrium manner. See section 3.1.1 for an example involving two competing size-structured species.

The analysis above, based upon the net reproductive rates n_1 and n_2, bears a striking similarity to the standard analysis of the famous Lotka–Volterra (unstructured) differential equation model for two species competition. The matrix M is analogous to the so-called community matrix in that classical theory. Unlike the classical theory, however, the criteria above can be related to class specific transition and fertility rates through the net reproductive numbers n_1 and n_2.

A host–parasitoid interaction. Parasitoids often attack a specific larval instar of their host. This fact can be of use in selecting parasitoids for a biological control program in which the host is a pest species. Given that a selection of parasitoids is available, which larval stage is optimally attacked if the goal is to reduce or eliminate the pest species? We will consider one of a variety of discrete age-structured host–parasitoid matrix models that were derived and studied by Barclay [22] with this question in mind.

Consider a multispecies system of the form (1.58) in which x is a vector of adult parasitoid life cycle stages and y is a vector of host life-cycle stages. The first $m_2 - 1$ components $y_1, \cdots, y_{m_2-1}$ are instars of *nonparasitized* larva densities and y_{m_2} is the adult (pest) host density. In this model $g(x, y) = Q(x, y)y$, where the host projection matrix is a nonlinear Leslie matrix of the form

$$Q(x,y) = \begin{bmatrix} 0 & 0 & \cdots & 0 & \cdots & 0 & f\exp(-gy_{m_2}) \\ \sigma & 0 & \cdots & 0 & \cdots & 0 & 0 \\ 0 & \sigma & \cdots & 0 & \cdots & 0 & 0 \\ \vdots & \vdots & \ddots & \vdots & \ddots & \vdots & \vdots \\ 0 & 0 & \cdots & \sigma\exp(-qp_1) & \cdots & 0 & 0 \\ \vdots & \vdots & \ddots & \vdots & \ddots & \vdots & \vdots \\ 0 & 0 & \cdots & 0 & \cdots & \sigma & 0 \end{bmatrix}.$$

The transition probability $\sigma < 1$ between age classes is taken to be independent of age, and the inherent per capita fertility rate is f. Note that adult fertility is density dependent (on adult population density only), as is expressed by the term $\exp(-gy_{m_2})$. Parasitoids of all stages are assumed to attack a single host instar, namely, the instar of age a, which is modeled by the term $\exp(-qp_1)$ located in the $(a+1)^{st}$ row and a^{th} column of the matrix $Q(x, y)$. In the absence of the parasitoids, the host population has a unique positive equilibrium

$$y_e = \frac{1}{g} \ln \left(f\sigma^{m_2 - 1} \right) \begin{bmatrix} \sigma^{1-m_2} \\ \sigma^{2-m_2} \\ \vdots \\ \sigma^{-1} \\ 1 \end{bmatrix}$$

provided $f\sigma^{m_2-1} > 1$. Let $y_a^e = \sigma^{a-m_2} \frac{1}{g} \ln(f\sigma^{m_2-1})$ denote the ath component of y_e. This equilibrium is stable if all eigenvalues ζ of the Jacobian $J_y g(0, y_e)$ satisfy $|\zeta| < 1$. These eigenvalues turn out to be $\zeta = (1 - \ln(f\sigma^{m_2-1}))^{1/m_2}$ and hence $y_e > 0$ is stable if

$$1 < f\sigma^{m_2-1} < e^2, \tag{1.62}$$

which we assume to hold.

The projection matrix of the parasitoid species has the form $T(x, y) + F(x, y)$, where the transition and fertility matrices are

$$T(x,y) = \begin{bmatrix} 0 & 0 & \cdots & 0 & 0 \\ s & 0 & \cdots & 0 & 0 \\ 0 & s & \cdots & 0 & 0 \\ \vdots & \vdots & \ddots & \vdots & \vdots \\ 0 & 0 & \cdots & s & 0 \end{bmatrix},$$

$$F(x,y) = qu^{\delta-k}\sigma^k \frac{1 - \exp(-qp_1)}{qp_1} y_a \begin{bmatrix} 1 & 1 & \cdots & 1 \\ 0 & 0 & \cdots & 0 \\ \vdots & \vdots & \ddots & \vdots \\ 0 & 0 & \cdots & 0 \end{bmatrix},$$

$$p_1 = \Sigma_{i=1}^{m_1} x_i, \quad S = \Sigma_{i=1}^{m_1} s^{i-1}.$$

The normalized fertility matrix and the inherent net reproductive number of the parasitoid (at the host equilibrium y_e) are

$$\Phi(x,y) = \frac{1}{S} \frac{1 - \exp(-qp_1)}{qp_1} \frac{y_a}{y_a^e} \begin{bmatrix} 1 & 1 & \cdots & 1 \\ 0 & 0 & \cdots & 0 \\ \vdots & \vdots & \ddots & \vdots \\ 0 & 0 & \cdots & 0 \end{bmatrix}, \quad n = qSu^{\delta-k}\sigma^k y_a^e.$$

In this parasitoid model it is assumed that the hosts do not reach the adult stage before mummification and emergence of the adult parasitoid. Prior to emergence parasitoid mortality is assumed the same as that as the host, which is σ for k time units, $k \leq \delta$ (the developmental time of the parasitoids), during which there is no change in host mortality due to parasitization, and $u \leq \sigma$ after k time units. After emergence, the adult parasitoid survival probability $s < 1$ is independent of stage and density.

In Theorems 1.5.1 and 1.5.2 with $\lambda = n$, $\lambda_0 = 1$ is a strictly dominant critical value with

$$v = \begin{bmatrix} 1 \\ s \\ \vdots \\ s^{m_1-1} \end{bmatrix} > 0, \quad w = \begin{bmatrix} \Sigma_{i=1}^{m_1} s^{i-1} \\ \Sigma_{i=1}^{m_1-1} s^{i-1} \\ \vdots \\ 1 \end{bmatrix} > 0,$$

and therefore

$$w^\tau Bv = w^\tau \Phi(0, y_e)v = S > 0.$$

Moreover, a calculation shows that

$$\lambda_1 = \frac{1}{2} q^2 S u^{\delta-k} \sigma^{k+a-m_2-2} \left(2 - \ln\left(f\sigma^{m_2-1}\right)\right)$$

and hence $\lambda_1 > 0$ by (1.62). It follows from Theorems 1.5.1 and 1.5.2 that there is a stable bifurcation of positive equilibria from the extinction equilibrium $x = 0$, $y = y_e > 0$ as n increases through 1. Such an equilibrium represents a stable coexistence state in which the parasitoid survives and reduces the equilibrium level of the host pest.

Returning to the question posed above, we wish to determine which host class a should be attacked if the adult component $y_{m_2} = y_{m_2}(a)$ of the positive equilibrium is to be minimized, holding all other parameters fixed (including n). In absence of an explicit formula for the positive equilibrium, we can use the Liapunov/Schmidt parameterization in Theorem 1.5.1, mathematically treating a as a continuous variable, to find

$$\frac{dy_{m_2}}{da} = \left(-Sqg^{-1} \ln \sigma\right) \varepsilon + O\left(\varepsilon^2\right).$$

Thus, this derivative is positive for $\varepsilon > 0$ small, which shows y_{m_2} is minimized at $a = 0$. That is to say, the "optimal" strategy is to parasitize the youngest larval instar. These result agrees with the conclusions in [22].

The age-structured model above assumes that only one larval stage is parasitized and then asserts that to parasitize the youngest stage is optimal. However, should only one larval instar stage be parasitized for the minimization of the adult host equilibrium level y_{m_2}? That is to say, for a given parasitization effort, could a further reduction in y_{m_2} be achieved by an optimal distribution of parasitization stages? This question is studied in [80], where a host–parasitoid (Usher type) model is studied with a distribution of parasitization stages and it is proved that it is optimal to parasitize only one larval stage, namely, the stage that produces maximal emergence numbers of parasitoids.

A discrete size-structured model for the chemostat. A discrete model for size-structured microbial growth in a chemostat of the form (1.57) is derived and studied in [178], [386]. In that model the microbial population is structured by m_1 size classes and y is the limiting nutrient concentration (y is a scalar so $m_2 = 1$). The components of x are the biomasses of these size classes at time t. In (1.57)

$$(1.63)\quad \begin{aligned} P &= (1-E)\begin{bmatrix} 1-\frac{1}{\nu-1}u(y) & \cdots & 0 & \frac{\nu}{\nu-1}u(y) \\ \frac{\nu}{\nu-1}u(y) & \cdots & 0 & 0 \\ \vdots & \ddots & \vdots & \vdots \\ 0 & \cdots & 1-\frac{1}{\nu-1}u(y) & 0 \\ 0 & \cdots & \frac{\nu}{\nu-1}u(y) & 1-\frac{1}{\nu-1}u(y) \end{bmatrix}, \\ g &= (1-E)\left(y - u(y)\sum_{i=1}^{m_1} x_i\right) + Ey_e, \end{aligned}$$

where $\nu = 2^{1/m_1}$, $E \in (0,1)$ is the "washout rate," $y_e > 0$ is the nutrient concentration supplied to the chemostat, and $u(y) \geq 0$ is the nutrient uptake rate per unit biomass per unit time (which, because it is assumed that biomass is measured in nutrient-equivalent units, is also the rate of increase in biomass per unit time per unit biomass). Assume that

$$u(y_e) > 0, \quad u'(y_e) > 0.$$

The prototypical nutrient uptake rate u is the Michaelis/Menten function

$$(1.64)\qquad u(y) = u_{\max}\frac{y}{a+y},$$

where $u_{\max}$ is the maximum uptake rate and $a > 0$ is the "Michaelis–Menten" or "half saturation" constant. The model parameters are required to satisfy certain constraints in order to make the model meaningful (e.g., so that the nutrient uptake in one time step does not exceed what is available). One such constraint is, for example, $u_{\max} < \nu - 1$. Others are more complicated and will not concern us here. See [178], [386] for more modeling details.

Take $\lambda = 1 - E$ and the model equations become

$$\begin{aligned} x(t+1) &= P(\lambda, y)x, \\ y(t+1) &= g(\lambda, x, y), \end{aligned}$$

where

$$P(\lambda, y) = \lambda \begin{bmatrix} 1-\frac{1}{\nu-1}u(y) & \cdots & 0 & \frac{\nu}{\nu-1}u(y) \\ \frac{\nu}{\nu-1}u(y) & \cdots & 0 & 0 \\ \vdots & \ddots & \vdots & \vdots \\ 0 & \cdots & 1-\frac{1}{\nu-1}u(y) & 0 \\ 0 & \cdots & \frac{\nu}{\nu-1}u(y) & 1-\frac{1}{\nu-1}u(y) \end{bmatrix},$$

$$g(\lambda, x, y) = \lambda \left(y - u(y) \sum_{i=1}^{m_1} x_i \right) + (1-\lambda)\, y_e.$$

Note that the model has the extinction equilibrium $x = 0$, $y = y_e$ for all $\lambda \in (0,1)$. Referring to the remark at the end of section 1.5.1 we apply Theorems 1.5.1 and 1.5.2 with $A = 0$ and

$$B = \begin{bmatrix} 1-\frac{1}{\nu-1}u(y_e) & \cdots & 0 & \frac{\nu}{\nu-1}u(y_e) \\ \frac{\nu}{\nu-1}u(y_e) & \cdots & 0 & 0 \\ \vdots & \ddots & \vdots & \vdots \\ 0 & \cdots & 1-\frac{1}{\nu-1}u(y_e) & 0 \\ 0 & \cdots & \frac{\nu}{\nu-1}u(y_e) & 1-\frac{1}{\nu-1}u(y_e) \end{bmatrix}.$$

Then

$$\lambda_0 = \frac{1}{1+u(y_e)} \in (0,1)$$

is a strictly dominant critical value of $A + \lambda B = \lambda B$ with eigenvectors

$$v = w = \begin{bmatrix} 1 \\ 1 \\ \vdots \\ 1 \end{bmatrix} > 0,$$

and hence $w^\tau B v = m_1 (1 + u(y_e)) > 0$. Since $J_y g(\lambda_0, 0, y_e) = 1 - \lambda_0 \neq 0$, Theorem 1.5.1 implies the existence of a continuum of positive equilibria bifurcating from the extinction equilibrium at $\lambda = \lambda_0$, i.e., at $E = E_0 \doteq \frac{u(y_e)}{1+u(y_e)} \in (0,1)$. Theorem 1.5.2 implies that the extinction equilibrium loses stability as λ increases through λ_0 (or as the washout rate E decreases through E_0). A calculation yields

$$\lambda_1 = -\frac{w^\tau [d_{ij}]\, v}{w^\tau B v} = m_1 \frac{u(y_e) u'(y_e)}{(1+u(y_e))^2} > 0,$$

and hence the bifurcation is stable. (Note that $u'(y_e) < 0$ would imply an unstable bifurcation.)

We have shown that for $E > E_0$ ($E < E_0$) the extinction equilibrium is locally asymptotically stable (unstable) and for $E < E_0$, but close to E_0, the

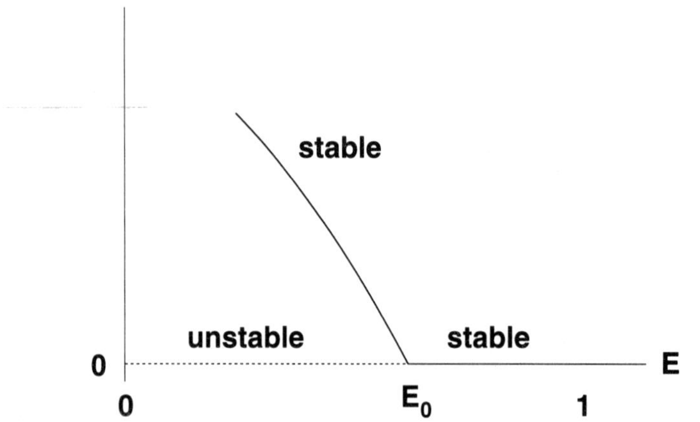

FIG. 1.23. *The bifurcation diagram for the discrete chemostat model* (1.57)–(1.63) *is schematically represented by a plot of the distance between the positive equilibrium pairs and the extinction equilibrium pair* $(0, E_0)$ *against the parameter* E.

positive equilibrium is locally asymptotically stable (see Fig. 1.23). In [386] these results, under the appropriate constraints on the parameters, are shown to be global in that the positive equilibrium stability is global for all $E \in (0, E_0)$ and the extinction equilibrium stability is global for all $E \in (E_0, 1)$.

The competition of several microbial species in a chemostat will be considered in section 3.1.1.

CHAPTER 2

Continuous Models

The matrix equations studied in Chapter 1 have several advantages when used as models in population dynamics. Models are relatively easy to construct from life-cycle information about individuals. Numerical simulations are particularly easy to carry out. Also, matrix models lend themselves to a description of general structuring variables and they are often (but certainly not always) more analytically tractable than continuous models.

Matrix models are best applied to populations with distinct "stages" with respect to the structuring variable (age, body size, etc.); they are at a disadvantage when such stages are not present. Matrix models cannot, of course, describe the dynamics of a population that occur between the discrete time steps of the model and do not lend themselves to modeling based on instantaneous rates of change. For a discussion of discrete and continuous structured models, their advantages and disadvantages, see [408].

In this chapter we consider structured models continuous in both the structuring variable and time. The general theory of continuous structured population dynamics involves partial differential equations and therefore is mathematically more difficult than that of matrix models. Even the most basic of questions, such as the existence, uniqueness, and positivity of solutions of initial value problems, involve formidable mathematical difficulties and technicalities. Moreover, the partial differential equation problems that arise are usually not of a classical type in that they are "nonlocal" (i.e., they are integro-partial differential equations) and may contain nonlinear, nonlocal boundary conditions. For these reasons, we restrict our attention to the mathematically simplest and most well-developed class of continuous models, namely, age-structured models. Moreover, we focus on asymptotic dynamics (specifically, equilibria and stability) and do not dwell on the mathematical theory of initial value problems (see [429], [244]).

We begin by deriving equations for continuous age-structured populations in continuous time (often called the *McKendrick equations* [318]) from the discrete Leslie matrix model by the limiting process of infinitesimally refining the age classes and the census time interval. (For other derivations see [236], [323].) We then give an equilibrium theory based on a bifurcation theoretic approach that closely parallels that given in Chapter 1 for matrix models. (Indeed, the same finite-dimensional Theorem B.1.1 will be used.) Applications illustrate the theory and the two basic types of bifurcations that can occur.

In this chapter we work directly with the partial differential equation form of the McKendrick model. Often the McKendrick model equations are used instead as a starting point for the derivation of other, more tractable, types of model equations, such as ordinary or delay differential equations. This is done by taking advantage of special mathematical features that derive from special modeling assumptions. Some examples of this approach will be studied in Chapter 3.

2.1 Age-structured models

In the classical Leslie matrix model for age-structured populations the projection time interval and the length of the age classes are equal and, for mathematical convenience, taken to be the unit of time. If the projection time interval and the length of the age classes are instead both equal to $h > 0$, then the Leslie matrix model takes the form

$$\begin{bmatrix} x_h(t+h) \\ x_{2h}(t+h) \\ \vdots \\ \vdots \\ x_{mh}(t+h) \end{bmatrix} = \begin{bmatrix} f_h(h) & f_{2h}(h) & \cdots & f_{(m-1)h}(h) & f_{mh}(h) \\ t_h(h) & 0 & \cdots & 0 & 0 \\ 0 & t_{2h}(h) & \cdots & 0 & 0 \\ \vdots & \vdots & \ddots & \vdots & \vdots \\ 0 & 0 & \cdots & t_{(m-1)h}(h) & 0 \end{bmatrix} \begin{bmatrix} x_h(t) \\ x_{2h}(t) \\ \vdots \\ \vdots \\ x_m(t) \end{bmatrix}, \tag{2.1}$$

where $x_a(t)$ denotes the number of individuals whose ages are between $a - h$ and a at time t. Here $mh = a_M$ is the maximal possible age of an individual and

$$f_a(h) = \begin{cases} \text{per capita number of (surviving) offspring produced during a} \\ \text{time interval of length } h \text{ by individuals of ages } a-h \text{ to } a, \end{cases}$$

$$t_a(h) = \begin{cases} \text{the fraction of those individuals of ages} \\ a-h \text{ to } a \text{ who survive } h \text{ units of time.} \end{cases}$$

Thus, $1 - t_a(h)$ is the fraction of those individuals of ages $a - h$ to a who do not survive h units of time. We make the following assumptions concerning these vital rates:

$$\begin{aligned} &t_a(0) = 1, \\ &\lim_{h\to 0+} \tfrac{1-t_a(h)}{h} = \delta(a), \ a \geq 0, \\ &\lim_{h\to 0+} \tfrac{f_a(h)}{h} = \beta(a), \ a \geq 0. \end{aligned}$$

Finally, we assume that a *density function* $\rho(t, a)$ exists (with respect to age a), defined for $a,\, t \geq 0$, such that

$$x_a(t) = \int_{a-h}^{a} \rho(t, \alpha) d\alpha. \tag{2.2}$$

That is to say, we assume that the population can be described by a function $\rho(t, a)$ such that the number of individuals of ages between a_1 and a_2 is given by $\int_{a_1}^{a_2} \rho(t, a) da$. Our goal is to derive an equation for $\rho(t, a)$ by passing $h \to 0$ in the Leslie matrix model (2.1).

Consider a fixed age $a = ih > 0$. From the Leslie model with projection matrix (2.1) we have

$$x_{a+h}(t + h) = t_a(h) x_a(t), \qquad a = h,\ 2h, \dots,\ (m-1)h.$$

From (2.2) these equations become

$$\int_a^{a+h} \rho(t + h, \alpha) d\alpha = t_a(h) \int_{a-h}^{a} \rho(t, \alpha) d\alpha$$

or

$$\int_{a-h}^{a} \rho(t + h, \alpha + h) d\alpha = t_a(h) \int_{a-h}^{a} \rho(t, \alpha) d\alpha.$$

Subtracting $\int_{a-h}^{a} \rho(t, \alpha) d\alpha$ from both sides and dividing the result by h^2, we obtain

$$\begin{aligned} &\tfrac{1}{h} \int_{a-h}^{a} \left[\tfrac{\rho(t+h, \alpha+h) - \rho(t, \alpha)}{h} \right] d\alpha \\ &\qquad = - \left[\tfrac{1 - t_a(h)}{h} \right] \tfrac{1}{h} \int_{a-h}^{a} \rho(t, \alpha) d\alpha. \end{aligned} \tag{2.3}$$

In order to pass to the limit $h \to 0$ we would need to make technical assumptions about the density function $\rho(t, a)$. We will not concern ourselves here with these technicalities, but simply proceed heuristically. Assume that the directional derivative

$$D\rho(t, a) \doteq \lim_{h \to 0} \frac{\rho(t + h, a + h) - \rho(t, a)}{h}$$

exists for $t, a > 0$. Then (formally) letting $h \to 0$ in (2.3) we obtain the equation

$$D\rho(t, a) = -\delta(a) \rho(t, a), \quad t > 0, \quad a > 0.$$

If $\rho(t, a)$ has partial derivatives at (t, a), this equation can be written

$$\partial_t \rho(t, a) + \partial_a \rho(t, a) = -\delta(a) \rho(t, a), \quad t > 0, \quad a > 0.$$

This is the most commonly used form of the differential equation for ρ, and we will use it here (although for rigorous treatments of the equation the directional derivative, or some "weak" formulation of the differential equation, is used).

From the nonnewborn classes $a > 0$ of the Leslie matrix model (2.1) we have heuristically arrived at a first-order partial differential equation for the density function $\rho(t,a)$. For the newborn class $a = 0$ we have from (2.1) for $t > 0$

$$x_h(t+h) = \sum_{j=1}^{m} f_{jh}(h) x_{jh}(t)$$

and

$$\frac{1}{h} x_h(t+h) = \sum_{j=1}^{m} \left[\frac{1}{h} f_{jh}(h)\right] \left[\frac{1}{h} x_{jh}(t)\right] h. \tag{2.4}$$

Now

$$\frac{1}{h} x_{jh}(t) = \frac{1}{h} \int_{(j-1)h}^{jh} \rho(t,\alpha) d\alpha,$$

and if we hold $jh = \alpha$ fixed and let $h \to 0$, we have (formally)

$$\lim_{h\to 0} \frac{1}{h} x_{jh}(t) = \rho(t,\alpha), \qquad \lim_{h\to 0} \frac{1}{h} x_h(t+h) = \rho(t,0).$$

Thus, (2.4) yields in the limit as $h \to 0$ the equation

$$\rho(t,0) = \int_0^{a_M} \beta(\alpha)\rho(t,\alpha)d\alpha, \qquad t > 0.$$

The equations for the density $\rho(t,a)$ of an age-structured population, together with an initial condition, are

$$\begin{array}{ll} \text{(a)} & \partial_t \rho(t,a) + \partial_a \rho(t,a) = -\delta\rho(t,a), \quad t > 0, \quad 0 < a < a_M \le +\infty, \\ \text{(b)} & \rho(t,0) = \int_0^{a_M} \beta\rho(t,a)da, \quad t > 0, \\ \text{(c)} & \rho(0,a) = \rho_0(a), \quad a \ge 0. \end{array} \tag{2.5}$$

If the maximal age is finite, $a_M < +\infty$, it is natural to require the additional boundary condition

$$\rho(t,a) = 0, \quad t \ge 0, \quad a \ge a_M.$$

In many treatments of continuous age-structured models it is assumed that there is no maximal age, i.e., $a_M = +\infty$. In this case, this boundary condition is replaced by some kind of control on $\rho(t,a)$ for large a (e.g., $\rho(t,\cdot) \in L^1$).

The vital death and fertility rates δ and β are allowed to depend explicitly on time t and on $\rho(t,a)$. In the latter case, the partial differential equation and/or the integral boundary birth condition are nonlinear. In population dynamics a reasonable assumption is that these vital rates depend on one or more functionals (weighted total population sizes) of the form $\int_0^{a_M} \omega(a)\rho(t,a)da$. An example is the total population size $\int_0^{a_M} \rho(t,a)da$. In this case the *McKendrick model*

(2.5) involves a nonlinear hyperbolic partial integro-differential equation with a nonlinear integral boundary condition.

Uribe [409] has proved that the Leslie matrix model, in the derivation above, is a consistent and stable discretization of the McKendrick equations. He makes rigorous the derivation of (2.5) above by showing that the Leslie model gives rise to a solution that converges to a (weak) solution of the McKendrick model.

Rigorous treatments of the basic theory of the age-structured McKendrick equations (2.5) can be found in [241], [429] (for $a_M = +\infty$), and [244] (for $a_M < +\infty$). Also see the seminal paper of Gurtin and MacCamy [198]. (See [248] and [404] for treatments of the size-structured McKendrick model.) These theories establish the well posedness of the initial/boundary value problem above, i.e., give conditions under which solutions exist and are unique. As is to be expected, these conditions involve appropriate Lipschitz conditions on the vital rates as functions of the density ρ. (Note that these difficult technicalities of existence and uniqueness do not arise for discrete matrix models, since as recursive formulas they obviously define unique solutions.) Moreover, these references establish the basic linearization principle for the study of stability of equilibria, either by nonlinear semigroup theory or by classical analysis.

2.2 Autonomous age-structured models

If the vital rates δ and β do not depend explicitly on time t, then the McKendrick model (2.5) is autonomous. We consider an autonomous McKendrick model in which δ and β depend on a finite number of weighted total population sizes and a finite maximal age

$$(2.6)\quad \begin{array}{ll} \text{(a)} & \partial_t \rho + \partial_a \rho = -\delta\left(a, p_1, \ldots, p_l\right) \rho, \ t > 0, \ 0 < a < a_M < +\infty, \\ \text{(b)} & \rho(t, 0) = \int_0^{a_M} \beta\left(a, p_1, \ldots, p_l\right) \rho(t, a) da, \ t > 0, \\ \text{(c)} & \rho(0, a) = \rho_0(a), \ 0 < a < a_M, \\ \text{(d)} & \rho(t, a_M) = 0, \ t > 0. \end{array}$$

Here the $p_i = p_i(t)$ are l functionals (weighted total population sizes) of the form

$$p_i(t) = \int_0^{a_M} \omega_i(a) \rho(t, a) da$$

with weights $0 \leq \omega_i(a) \in L^1[0, a_M]$. We are interested in asymptotic dynamics, and we will begin by studying the existence and stability of equilibrium solutions.

Conditions sufficient to guarantee the existence of a unique local solution $\rho(\cdot, a) \in C([0, T], L^1(0, a_M))$, for an initial density $0 \leq \rho_0(a) \in L^1(0, a_M)$, are given in [244, p. 51], namely, for $(p_1, \dots, p_l) \in \Omega$ and $a \in [0, a_M]$

$$
(2.7)\qquad
\begin{aligned}
&\delta(\cdot, p_1, \dots, p_l) \in L^1_{loc}([0, a_M), R_+),\\
&\beta(\cdot, p_1, \dots, p_l) \in L^1((0, a_M), R_+),\\
&0 \le \beta(a, p_1, \dots, p_l) \le \beta_l < +\infty,\ 0 \le \delta(a, p_1, \dots, p_l),\\
&\delta \text{ and } \beta \text{ are locally Lipschitz continuous with}\\
&\text{respect to each } p_i \text{ uniformly for } a \in [0, a_M],\\
&\textstyle\int_0^{a_M} \delta(a, p_1, \dots, p_l)\, da = +\infty.
\end{aligned}
$$

Here Ω is an open neighborhood of R^l_+. (A solution has a directional derivative $D\rho$ in place of $\partial_t \rho + \partial_a \rho$.) We assume in addition that for some integer $k \in I[0, +\infty)$ and each $a \in [0, a_M]$

$$
(2.8)\qquad \delta(a, \cdot, \dots, \cdot),\ \beta(a, \cdot, \dots, \cdot) \in C^{k+1}(\Omega, R_+).
$$

2.2.1 The extinction equilibrium. The McKendrick model (2.6) has the extinction equilibrium $\rho = 0$. The linearization of (2.6) at this equilibrium is

$$
(2.9)\qquad
\begin{aligned}
&\text{(a)}\quad \partial_t \rho + \partial_a \rho = -\delta(a)\rho, \quad t > 0, \quad 0 < a < a_M < +\infty,\\
&\text{(b)}\quad \rho(t, 0) = \textstyle\int_0^{a_M} \beta(a)\rho(t, a)da, \quad t > 0,\\
&\text{(c)}\quad \rho(0, a) = \rho_0(a), \quad 0 < a < a_M,\\
&\text{(d)}\quad \rho(t, a_M) = 0, \quad t > 0,
\end{aligned}
$$

where $\delta(a) \doteq \delta(a, 0, \dots 0)$ and $\beta(a) \doteq \beta(a, 0, \dots 0)$. Treatments of this linear McKendrick model can be found in [236] for $a_M = +\infty$ and [244] for $a_M < +\infty$. We will briefly outline the latter treatment.

The probability of living to age a is

$$
\Pi(a) \doteq \exp\left(-\int_0^a \delta(\alpha)d\alpha\right).
$$

Letting $\psi(t, a) = \rho(t, a)/\Pi(a)$, we find from (2.9(a)) that ψ satisfies the partial differential equation $\partial_t \psi + \partial_a \psi = 0$. The "general" solution of this equation is $\psi = \varphi(t - a)$, where φ is an arbitrary function. Thus, $\rho(t, a) = \varphi(t - a)\Pi(a)$, where φ is a function to be determined by the boundary and initial conditions (2.9(b),(c)). The boundary condition (2.9(d)) is satisfied by the assumption on δ in (2.7). The initial condition (2.9(c)) is satisfied by defining $\varphi(t)$ for negative t by $\varphi(t) = \rho_0(-t)/\Pi(-t)$, where $\rho_0(a)$ is extended to be 0 for $a > a_M$. Finally, for $t \ge 0$, $\varphi(t)$ is determined by the birth condition (2.9(b)). With $\beta(a)$ and $\Pi(a)$ extended to be 0 for $a > a_M$, the equation for $\varphi(t)$ obtained from (2.9(b)) is

$$
(2.10)\qquad \varphi(t) = \varphi_0(t) + \int_0^t \kappa(t - a)\varphi(a)da, \quad t \ge 0,
$$

where for $t \geq 0$

$$\varphi_0(t) \doteq \int_t^{+\infty} \beta(a)\rho_0(a-t)\frac{\Pi(a)}{\Pi(a-t)}da \geq 0, \quad \kappa(t) \doteq \beta(t)\Pi(t) \geq 0.$$

Equation (2.10) is a Volterra integral equation for $\varphi(t)$. The relationship between the solution of this integral equation and the solution of the linear McKendrick model (2.9) is made mathematically rigorous in [244, Chapters I and II]. A solution of (2.10) yields a solution of (2.9)

$$\rho(t,a) = \begin{cases} \rho_0(a-t)\frac{\Pi(a)}{\Pi(a-t)}, & 0 < t < a, \\ \varphi(t-a)\Pi(a), & a < t \end{cases}$$

and vice versa. (Here we ignore the mathematical technicalities in defining a solution of (2.9).) Equation (2.10) can be studied using Laplace transforms. The transform of the solution is given by $\widehat{\varphi}(z) = \widehat{\varphi}_0(z)/(1-\widehat{\kappa}(z))$. Consider the "characteristic equation" $\widehat{\kappa}(z) = 1$, i.e., the equation

$$\int_0^{+\infty} \beta(a)\Pi(a)e^{-za}da = 1 \tag{2.11}$$

for complex roots z. With $z = x$ restricted to real numbers, from assumptions (2.7) it follows that $\widehat{\kappa}(x) \geq 0$ is strictly decreasing for $x \in (-\infty, +\infty)$ with $\lim_{x\to-\infty} \widehat{\kappa}(x) = +\infty$ and $\lim_{x\to+\infty} \widehat{\kappa}(x) = 0$. Thus, the characteristic equation (2.11) has a unique real root $x = r$. If $z = x + iy$ is any other root,

$$\begin{aligned} \int_0^{+\infty} \beta(a)\Pi(a)e^{-ra}da &= 1 \\ &= \int_0^{+\infty} \beta(a)\Pi(a)e^{-xa}\cos(ya)\,da \\ &< \int_0^{+\infty} \beta(a)\Pi(a)e^{-xa}da, \end{aligned}$$

which implies that $r > x = \operatorname{Re} z$. Thus, there exists a strictly dominant real root of the characteristic equation (which is analogous to the strictly dominant eigenvalue of a projection matrix and hence the inherent growth rate for discrete matrix models). It is shown in [244] (also see [236] for the same result when $a_M = +\infty$) that the solution of (2.10) has the form

$$\varphi(t) = ce^{rt}\left(1+\eta(t)\right), \quad c \doteq \frac{\int_0^{+\infty}\varphi_0(t)e^{-rt}dt}{\int_0^{+\infty}t\kappa(t)e^{-rt}dt}, \quad \lim_{t\to+\infty}\eta(t) = 0.$$

Thus, the solution of the linear McKendrick (2.9) is

$$\rho(t,a) = \begin{cases} \rho_0(a-t)\frac{\Pi(a)}{\Pi(a-t)}, & 0 < t < a, \\ ce^{r(t-a)}\left(1+\eta(t-a)\right)\Pi(a), & a < t. \end{cases} \tag{2.12}$$

The asymptotic dynamics of $\rho(t,a)$ are determine by the dominant root r. Note that there is a constant $\xi > 0$ such that $p(t) \leq \xi e^{rt} p(0)$, where

$$p(t) = \int_0^{a_M} |\rho(t,a)|\, da \geq 0$$

is the total population size at time t. If $r < 0$, then for each $\varepsilon > 0$, $p(0) < \delta = \varepsilon/\xi$ implies that $p(t) < \varepsilon$ for all $t \geq 0$ and $\lim_{t\to+\infty} p(t) = 0$. In this sense, the extinction equilibrium is asymptotically stable if $r < 0$. On the other hand, if $r > 0$, then $p(t)$ is exponentially unbounded as $t \to +\infty$ and the extinction equilibrium is unstable. Thus, the extinction equilibrium loses stability as r increases through the critical value 0. Moreover, if $r = 0$, then there exist infinitely many equilibrium solutions of (2.9) given by $\rho(t,a) = k\Pi(a)$, where k is an arbitrary constant.

Since $\Pi(a)$ is the probability of living to age a, $\beta(a)\Pi(a)$ is the number of offspring per individual given that reaches age a. Thus,

$$n \doteq \widehat{\kappa}(0) = \int_0^{a_M} \beta(a)\Pi(a)da$$

is the expected number of offspring per individual per lifetime, which we define to be the *inherent net reproductive number.* Since $\widehat{\kappa}(x)$ is strictly decreasing to 0 as $x \to +\infty$, we see from the characteristic equation (2.11) that $r < 0$ if (and only if) $n < 1$, $r > 0$ if (and only if) $n > 1$ and $r = 0$ if (and only if) $n = 1$. For the McKendrick age-structured model this is the equivalent of Theorem 1.1.3 for matrix equations.

From (2.12) for $t > a_M$ we have

$$\lim_{t\to+\infty} \frac{\rho(t,a)}{p(t)} = \lim_{t\to+\infty} \frac{ce^{r(t-a)}\left(1+\eta(t-a)\right)\Pi(a)}{ce^{rt}\int_0^{a_M} e^{-ra}\left(1+\eta(t-a)\right)\Pi(a)da}$$

and hence that the *normalized age distribution* of any nontrivial solution ($c \neq 0$) satisfies

$$\lim_{t\to+\infty} \frac{\rho(t,a)}{p(t)} = \frac{e^{-ra}\Pi(a)}{\int_0^{a_M} e^{-ra}\Pi(a)da}. \tag{2.13}$$

Thus, the normalized distribution obtained from all solutions tends to the same asymptotic limit, called the *stable age distribution.* This property is analogous to property (1.7) for general linear matrix models. Furthermore, an integration of (2.9) from $a = 0$ to a_M yields

$$\begin{aligned} p'(t) &= \int_0^{a_M} \left(\beta(a) - \delta(a)\right)\rho(t,a)da \\ &= \left[\int_0^{a_M} \left(\beta(a) - \delta(a)\right)\frac{\rho(t,a)}{p(t)}da\right] p(t). \end{aligned}$$

Replacing the normalized distribution by its limit (2.13) we obtain the *limiting equation*

$$q'(t) = \left[\int_0^{a_M} (\beta(a) - \delta(a)) \frac{e^{-ra}\Pi(a)}{\int_0^{a_M} e^{-ra}\Pi(a)da} da \right] q(t).$$

Observing that

$$\begin{aligned}\int_0^{a_M} [\beta(a) - \delta(a)] \, e^{-ra}\Pi(a)da &= 1 + \int_0^{a_M} e^{-ra}\Pi'(a)da \\ &= 1 + e^{-ra}\Pi(a)\Big|_{a=0}^{a_M} + r \int_0^{a_M} e^{-ra}\Pi(a)da \\ &= r \int_0^{a_M} e^{-ra}\Pi(a)da\end{aligned}$$

we see that the limiting equation reduces to

$$q'(t) = rq(t). \tag{2.14}$$

This scalar ordinary differential equation correctly predicts the asymptotic dynamics of the population level quantity $p(t)$.

From the above results we see that the theory of linear age-structured McKendrick equations (2.9) is analogous to the theory of linear matrix equations given in section 1.1.2. In the following section we pursue this analogy with regard to equilibrium theory for nonlinear age-structured McKendrick equations (2.6).

2.2.2 Positive equilibria. In analogy to the general bifurcation results for matrix equations in section 1.2.3 we anticipate that positive equilibria of the general McKendrick model (2.6) will bifurcate from the extinction equilibrium as a parameter is varied in such a way as to cause the dominant root r to change sign (or equivalently to cause the inherent net reproductive number n to pass through 1). We will prove that this is in fact true by using Theorem B.1.1 in Appendix B.1; the same theorem we used for the matrix model analysis in Chapter 1.

With

$$\beta(a, p_1, \dots, p_l) \doteq nb(a, p_1, \dots, p_l),$$

where b is normalized to satisfy

$$\int_0^{a_M} b(a, 0, \dots, 0)\Pi(a)da = 1, \tag{2.15}$$

n is the inherent net reproductive number. Introducing this expression for β into (2.6) we obtain the equilibrium equations

$$\begin{array}{ll}\text{(a)} & \partial_a \rho(a) = -\delta(a, p_1, \dots, p_l)\rho(a), \quad 0 < a < a_M < +\infty, \\ \text{(b)} & \rho(0) = n \int_0^{a_M} b(a, p_1, \dots, p_l)\rho(a)da, \\ \text{(c)} & \rho(a_M) = 0, \\ \text{(d)} & p_i = \int_0^{a_M} \omega_i(a)\rho(a)da.\end{array} \tag{2.16}$$

From (a) and (c) we have

$$\rho(a) = c\Pi(a, p_1, \dots, p_l), \tag{2.17}$$
$$\Pi(a, p_1, \dots, p_l) \doteq \exp\left(-\int_0^a \delta(\alpha, p_1, \dots, p_l)d\alpha\right).$$

Note that $\Pi(a) = \Pi(a, 0, \dots, 0)$. Placing this expression into (b) and (d) we obtain the $l+1$ algebraic equations

$$c = nc\int_0^{a_M} b(a, p_1, \dots, p_l)\Pi(a, p_1, \dots, p_l)da, \tag{2.18}$$
$$p_i = c\int_0^{a_M} \omega_i(a)\Pi(a, p_1, \dots, p_l)da, \quad i = 1, \dots, l$$

for the $l+1$ unknowns c, $p_1, \dots, p_l$. From a solution of these algebraic equations we define a function $\rho(a)$ by (2.17), which by (2.7) is continuous for $a \in [0, a_M]$. This function solves the original equilibrium equations at points where it is differentiable (a further smoothness condition that depends on the smoothness of δ as a function of a). We define an *equilibrium of* (2.6) to be a function given by (2.17) and (2.18).

The system of equations (2.18) is equivalent to the system obtained by substituting the right-hand side of the first equation for c in the second equation. The resulting system can be written

$$\begin{aligned} c &= nc + nch_1(p_1, \dots, p_l), \\ p_i &= n\pi_i^0 c + nch_i(n, p_1, \dots, p_l), \end{aligned}$$

where

$$\pi_i^0 \doteq \int_0^{a_M} \omega_i(a)\Pi(a)da$$

and $h_i \in C^{k+1}(\Omega, R_+)$ satisfy $|h_i(p_1, \dots, p_l)| = O(\Sigma_{i=1}^l |p_i|)$ near the origin. This system of algebraic equations for c and p_i have the form of the equation studied in Appendix B.1, namely,

$$x = nLx + h(n, x),$$

where

$$x = \begin{bmatrix} c \\ p_1 \\ \vdots \\ p_l \end{bmatrix}, \qquad L = \begin{bmatrix} 1 & 0 & \cdots & 0 \\ \pi_1^0 & 0 & \cdots & 0 \\ \vdots & \vdots & \ddots & \vdots \\ \pi_l^0 & 0 & \cdots & 0 \end{bmatrix},$$

$$|h(n, x)| = o(|x|) \text{ near } x = 0.$$

The only characteristic value of the $(l+1) \times (l+1)$ matrix L is $n = 1$ and it has the positive characteristic vector

$$v = \begin{bmatrix} 1 \\ \pi_1^0 \\ \vdots \\ \pi_l^0 \end{bmatrix} > 0.$$

By Theorem B.1.1 (see Appendix B.1) there exists a continuum $\mathfrak{C}^+$ of nonnegative equilibrium pairs (n, x) containing $(1, 0)$ for which $\mathfrak{C}^+ / \{R^1 \times \partial R^l_+\}$ is nonempty (and consists of positive equilibrium pairs) and for which the following alternatives hold : (i) either $\mathfrak{C}^+ / \{(1, 0)\}$ is unbounded in $R^1 \times R^l_+$ and contains only positive equilibrium pairs, or (ii) $\mathfrak{C}^+$ contains a boundary pair $(n^*, x^*) \in R^1 \times \partial R^l_+$ for which $x^* \neq 0$. The second alternative can be ruled out, however, since the only solution of (2.18) lying on the boundary ∂R^l_+ is $x^* = 0$. (To see this, note from (2.18) that $c = 0$ implies that all $p_i = 0$ and if at least one $p_i = 0$, then $c = 0$ and hence all other $p_i = 0$.) From the second equilibrium equation in (2.18), together with $0 \leq \Pi(a, p_1, \ldots p_l) \leq 1$, we see that if any one component in x is unbounded along $\mathfrak{C}^+$, then all components are unbounded, including c. The continuum $\mathfrak{C}^+$ therefore defines a continuum C^+ of equilibrium pairs $(n, \rho) \in R \times C^0[0, a_M]$, from (2.17), which is unbounded (using, for example, the norm $|n| + \max_{[0,a_M]} |\rho(a)|$).

These results are summarized in the following theorem.

THEOREM 2.2.1. *Under assumptions* (2.7) *and* (2.8), *the McKendrick model* (2.6) *has an unbounded continuum C^+ of equilibrium pairs (n, ρ) such that $(1, 0) \in C^+$ and $C^+ / \{(1, 0)\}$ consists only of positive equilibrium pairs.*

From the first equation in (2.18) we have, for any positive equilibrium,

$$n\nu(p_1, \ldots, p_l) = 1, \quad \nu(0, \ldots, 0) = 1, \tag{2.19}$$

where

$$\nu(p_1, \ldots, p_l) \doteq \int_0^{a_M} b(a, p_1, \ldots, p_l)\Pi(a, p_1, \ldots, p_l)da.$$

The quantity $n\nu(p_1, \ldots, p_l)$ is the *net reproductive number at equilibrium*, and (2.19) means that at equilibrium each individual exactly replaces itself. Mathematically, this equation can be used to determine properties of the bifurcating continuum of equilibria in the way the analogous equation (1.35) is used for matrix models in section 1.2.5. For example if b and δ are such that $\nu(p_1, \ldots, p_l) \to 0$ as $\Sigma_{i=1}^l |p_i| \to +\infty$, then by (2.19) the spectrum $\sigma(C^+) = \{n : (n, \rho) \in C^+\}$ associated with C^+ must be infinite. Since $0 \notin \sigma(C^+)$, the spectrum must be an interval of the form $\sigma(C^+) = (n_0, +\infty)$ or $[n_0, +\infty)$ for $0 < n_0 \leq 1$.

Properties of equilibria near the bifurcation point can be determined from the Liapunov/Schmidt parameterization of the branch $\mathfrak{C}^+$. See Theorem B.2.1 in Appendix B.2. If $k \geq 2$ in (2.8), then for $\varepsilon > 0$ small

$$n = 1 + n_1\varepsilon + O(\varepsilon^2), \quad c = \varepsilon + O(\varepsilon^2), \quad p_i = \pi_i^0\varepsilon + O(\varepsilon^2). \tag{2.20}$$

Substituting these expansions in (2.19) and differentiating the result with respect to ε, we obtain, upon setting $\varepsilon = 0$,

$$n_1 = \int_0^{a_M} \Pi(a) \left(b(a) \int_0^a D\delta(\alpha)d\alpha - Db(a) \right) da, \tag{2.21}$$

where we have defined

$$D\delta(a) \doteq \sum_{i=1}^{l} \pi_i^0 \partial_{p_i} \delta(a, 0, \ldots, 0), \quad Db(a) \doteq \sum_{i=1}^{l} \pi_i^0 \partial_{p_i} b(a, 0, \ldots, 0).$$

The bifurcation is to the right (to the left) if $n > 1$ (if $n < 1$) for equilibrium pairs from C^+ near the $(n, \rho) = (1, 0)$. It follows that the bifurcation is to the right if $n_1 > 0$ and is to the left if $n_1 < 0$.

We might expect a relationship between the direction of bifurcation and the stability of the equilibria from pairs near the equilibrium point, just as there is for nonlinear matrix equations. We consider the stability of the positive equilibria on the bifurcating continuum C^+ by a formal linearization procedure. For a rigorous proof of the validity of the linearization procedure see [244, p. 69]. (Also see [429] for the case $a_M = +\infty$.)

If $\rho_e(a)$ is an equilibrium, then the linearization of (2.6) with $\beta = nb$ is obtained by substituting $\rho(t, a) = z(t, a) + \rho_e(a)$ into the equations and dropping all nonlinear terms in the resulting equation for $z(t, a)$. This results in the equations

$$\begin{aligned} \partial_t z + \partial_a z &= -\delta(a, p_1^e, \ldots, p_l^e) z \\ &\quad - \rho_e(a) \sum_{i=1}^{l} \partial_{p_i} \delta(a, p_1^e, \ldots, p_l^e) \int_0^{a_M} \omega_i(\alpha) z(t, \alpha) d\alpha, \\ z(t, 0) &= n \int_0^{a_M} b(a, p_1^e, \ldots, p_l^e) z(t, a) da \\ &\quad + n \int_0^{a_M} \sum_{i=1}^{l} \partial_{p_i} b(a, p_1^e, \ldots, p_l^e) \rho_e(a) da \int_0^{a_M} \omega_i(\alpha) z(t, \alpha) d\alpha, \end{aligned}$$

where $p_i^e = \int_0^{a_M} \omega_i(a) \rho_e(a) da$. We look for solutions of the form $z(t, a) = y(a) \exp(-\zeta t)$. A substitution leads to the equations

$$\begin{aligned} y'(a) &= -(\zeta + \delta(a, p_1^e, \ldots, p_l^e)) y(a) \\ &\quad - \rho_e(a) \sum_{i=1}^{l} \partial_{p_i} \delta(a, p_1^e, \ldots, p_l^e) \int_0^{a_M} \omega_i(\alpha) y(\alpha) d\alpha, \\ y(0) &= n \int_0^{a_M} b(a, p_1^e, \ldots, p_l^e) y(a) da \\ &\quad + n \int_0^{a_M} \sum_{i=1}^{l} \partial_{p_i} b(a, p_1^e, \ldots, p_l^e) \rho_e(a) da \int_0^{a_M} \omega_i(\alpha) y(\alpha) d\alpha \end{aligned} \tag{2.22}$$

for $y(a)$. For the equilibria from C^+ we have from (2.20) and (2.17)

$$n = 1 + n_1 \varepsilon + O(\varepsilon^2), \quad \rho_e(a) = \Pi(a) \varepsilon + O(\varepsilon^2), \quad p_i^e = \pi_i^0 \varepsilon + O(\varepsilon^2), \tag{2.23}$$

and consequently y and ζ have expansions

$$y(a) = y_0(a) + y_1(a)\varepsilon + O(\varepsilon^2), \quad \zeta = \zeta_0 + \zeta_1\varepsilon + O(\varepsilon^2).$$

We know the extinction equilibrium loses stability at $\varepsilon = 0$ (i.e., $n = 1$) at which the dominant root of the characteristic equation (2.11) equals 0. Thus, $\zeta_0 = 0$ and we are interested in the sign of ζ for $\varepsilon > 0$, i.e., in the sign of the coefficient ζ_1. A substitution of these expansions into (2.22) leads, from the lowest-order terms in ε, to $y_0(a) = \Pi(a)$ and, from the first-order terms in ε, to the equations

$$\begin{aligned} y_1'(a) &= -\delta(a)y_1(a) - \Pi(a)\left(\zeta_1 + 2D\delta(a)\right), \\ y_1(0) &= \int_0^{a_M} b(a)y_1(a)da + n_1 + 2\int_0^{a_M} \Pi(a)Db(a)da. \end{aligned}$$

From the first equation (which is a scalar, linear nonhomogeneous first-order ordinary differential equation for y_1) we obtain

$$y_1(a) = y_1(0)\Pi(a) - \Pi(a)\int_0^a \left(\zeta_1 + 2D\delta(\alpha)\right) d\alpha,$$

which when substituted into the second equation yields

$$\begin{aligned} y_1(0) &= y_1(0) - \int_0^{a_M} b(a)\Pi(a)\int_0^a \left(\zeta_1 + 2D\delta(\alpha)\right) d\alpha da \\ &\quad + n_1 + 2\int_0^{a_M} \Pi(a)Db(a)da. \end{aligned}$$

Solving this equation for ζ_1 and using the formula (2.21), we obtain

$$\zeta_1 = -\frac{n_1}{\int_0^{a_M} ab(a)\Pi(a)da}.$$

This shows that the equilibria (2.23) are locally asymptotically stable if the bifurcation is to the right (i.e., if $n_1 > 0$).

Recall that the bifurcation of C^+ is called stable (unstable) if the positive equilibria from C^+ near the bifurcation point $(n, \rho) = (1, 0)$ are locally asymptotically stable (unstable).

THEOREM 2.2.2. *Assume* (2.7) *and* (2.8).

(a) *The extinction equilibrium* $\rho = 0$ *of the McKendrick model* (2.6) *is (locally asymptotically) stable if* $n < 1$ *and is unstable if* $n > 1$.

(b) *Assume that* $k \geq 2$ *and* $n_1 \neq 0$. *The bifurcation of* C^+ *is stable if it is to the right* ($n_1 > 0$) *and unstable if it is to the left* ($n_1 < 0$).

The expansions (2.23) can be used to study the nonlinear effects on the normalized age distribution near the bifurcation point. To do this the expansion for the equilibrium $\rho_e(a)$ must be carried out to order ε^2. This yields

$$\begin{aligned} \rho_e(a) &= \Pi(a)\varepsilon + \rho_1(a)\varepsilon + O(\varepsilon^3), \\ \rho_1(a) &= \rho_1(0)\Pi(a) - \int_0^a D\delta(\alpha)d\alpha, \\ \rho_1(0) &= \frac{\int_0^{a_M} \Pi^2(a)\int_0^a D\delta(\alpha)d\alpha da}{\int_0^{a_M} \Pi^2(a)da}. \end{aligned}$$

From $p_e = \int_0^{a_M} \rho_e(a)da$ and $d/d\varepsilon = (dn/d\varepsilon)(d/dn)$ we get

$$\begin{aligned}\frac{d}{dn}\left(\frac{\rho_e(a)}{p_e}\right) &= \frac{\int_0^{a_M}\left(\Pi(\alpha)\rho_1(a) - \Pi(a)\rho_1(\alpha)\right)d\alpha}{n_1\left(\int_0^{a_M}\Pi(a)da\right)^2} \\ &= \frac{\Pi(a)\int_{\alpha=0}^{a_M}\int_{s=a}^{\alpha} D\delta(s)ds\Pi(\alpha)d\alpha}{n_1\left(\int_0^{a_M}\Pi(a)da\right)^2}.\end{aligned}$$

For example, if nonlinear effects serve to increase the death rate of all ages ($D\delta(a) > 0$ for all a) and if the bifurcation is stable ($n_1 > 0$), then for $n \approx 1$

$$\begin{aligned}\frac{d}{dn}\left(\frac{\rho_e(a)}{p_e}\right) &> 0 \quad \text{for } a \approx 0, \\ \frac{d}{dn}\left(\frac{\rho_e(a)}{p_e}\right) &< 0 \quad \text{for } a \approx a_M,\end{aligned}$$

which shows that the equilibrium population gets younger as $n > 1$ increases.

For similar results with more general dependencies of δ and b on ρ see [88], [89].

Often the assumption of a finite maximal age $a_M < +\infty$ is not made and individuals of all ages are theoretically allowed. To consider McKendrick models with $a_M = +\infty$ one has to "control" ρ, b, and δ at $a = +\infty$ in some manner. For example, in place of the last assumption in (2.7) (which is needed for the boundary condition $\rho(t, a_M) = 0$) it is often required that $\rho \in L^1(0, +\infty)$; e.g., see [429]. Under these conditions Theorems 2.2.1 and 2.2.2 remain valid.

2.2.3 Hopf bifurcation. The stability properties of the positive equilibria near the bifurcation point described in Theorem 2.2.2 may not persist globally along the continuum C^+. Should the positive equilibria change stability as n is changed, due to a root ζ of the characteristic equation of the linearization at the equilibria crossing the imaginary axis, then (usually) another bifurcation occurs. If the root moves through $\zeta = 0$, the bifurcation involves equilibria, the possibilities being those discussed for discrete matrix models in section 1.2.3 (saddle-node, transcritical, and pitchfork). If, however, a complex conjugate pair crosses the imaginary axis when $n = n_{cr}$ through points $\zeta = \pm i\theta$, $\theta \neq 0$, and it does so transversely (i.e., $d\,\mathrm{Re}\,\zeta(n_{cr})/dn \neq 0$), then a classical Hopf bifurcation occurs in which time periodic solutions are created. These periodic solutions arise as small amplitude oscillations around the destabilized equilibrium with period approximately equal to $\theta/2\pi$. Their stability or instability can be determined, in principle, by known formulas, the use of which is unfortunately difficult in applications. Numerical investigations are commonly used to investigate Hopf bifurcations. A Hopf bifurcation theorem specifically for the McKendrick model (2.6) can be found in [87].

An illustration of a Hopf bifurcation to a limit cycle (followed by further bifurcations to chaos) can be seen in Figs. 2.1 and 2.2, where solutions of the

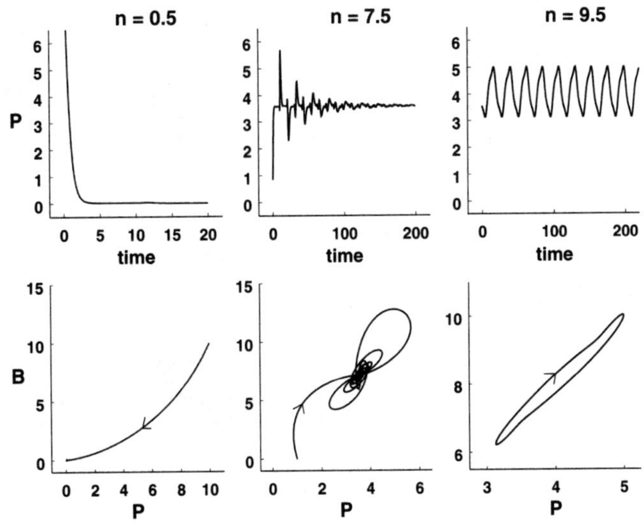

FIG. 2.1. *For the delay equations* (2.24) *the inherent net reproductive number is* $n = b\delta^{-1}$. *The population goes extinct for* $n = 0.5$, *equilibrates to a positive equilibrium for* $n = 7.5$, *and (after a Hopf bifurcation occurs) oscillates periodically for* $n = 9.5$. *Plots are shown for* $\delta = 2$, $c = 0.5$, *and* $T = 10$.

McKendrick model (2.6) are plotted for

$$\begin{aligned}\delta &= \text{constant} > 0,\\ \beta &= b\chi_T(a)\exp\left(-cp(t-T)\right), \quad b > 0, \quad c > 0.\end{aligned}$$

Here $\chi_T(a)$ is the Dirac function at $T > 0$, and hence in this model only individuals of age $a = T$ are fertile. The nonlinear density effects in this model occur in an individual's fertility rate, which is dependent on the total population size at birth. The model equations

$$\begin{aligned}&\partial_t\rho + \partial_a\rho = -\delta\rho,\\ &\rho(t,0) = bp(t-T)\exp\left(-cp(t-T)\right)\end{aligned}$$

lead to the system of delay differential equations

$$\begin{aligned}&p'(t) = -\delta p(t) + B(t),\\ &B'(t) = -(c+\delta)B(t) + bB(t-T)\exp\left(-cp(t-T)\right)\end{aligned} \tag{2.24}$$

for total population size $p(t)$ and the *total birth rate* $B(t) \doteq \rho(t,0)$.

2.3 Some applications

We look briefly at two applications that involve two fundamentally different types of nonlinear interactions in population dynamics: "negative feedback" (or deleterious) density effects and advantageous density effects (or Allee effects).

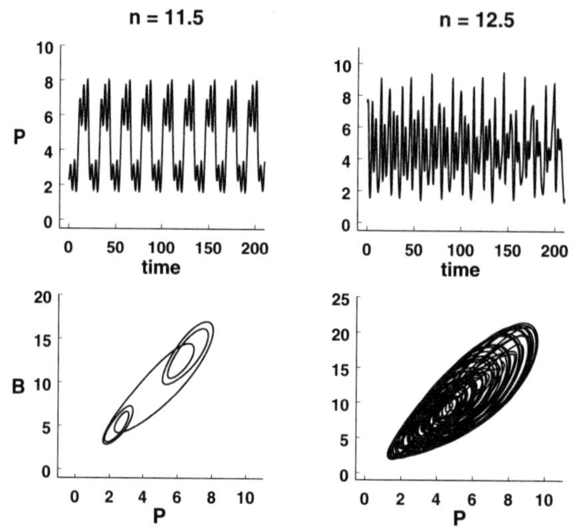

FIG. 2.2. *With further increases in n, the periodic solution in Fig. 2.1 becomes more complicated for $n = 11.5$ and an apparent chaotic oscillation occurs for $n = 12.5$.*

Negative feedback effects of density lead to stable bifurcations while Allee effects usually lead to unstable bifurcations. In the first application we consider a McKendrick model which allows for competition for a limited resource between juveniles and adults. The competition is viewed as deleterious in the sense that increased population levels result in decreased fertility and/or death rates. The second application involves an Allee effect [5], [130] and an unstable bifurcation. In this application unstable equilibria on the bifurcating continuum are stabilized at a saddle-node bifurcation at a critical value $n = n_{cr} < 1$. This is often the case in population models with an Allee effect at low population densities but negative feedback effects at high population densities.

Juvenile versus adult competition. Suppose that individuals in a population are nonreproducing prior to reaching a maturation age (or period) $a_j < a_M$. Then in the McKendrick model equations (2.5) $b = 0$ for all ages $a \in [0, a_j]$ and $b > 0$ for $a \in (a_j, a_M]$. Individuals of ages 0 to a_j are called "juveniles" while those of ages greater than a_j are called "adults." We are interested in competitive interactions between juveniles and adults and how this competition effects the dynamics of the population. To do this the birth and death rates are assumed age specific and dependent on weighted total numbers of juveniles and/or adults

$$p_J(t) = \int_0^{a_j} \omega_J(\alpha)\rho(t,\alpha)d\alpha, \quad p_A(t) = \int_{a_j}^{a_M} \omega_A(\alpha)\rho(t,\alpha)d\alpha$$

for nonnegative weights ω_J and ω_A. For weights $\omega_J \equiv \omega_A \equiv 1$ p_J and p_A reduce to total juvenile and adult numbers, respectively. These more general weighted class sizes allow differential effects of age classes on vital rates.

We consider the case when only the birth rate b is density dependent and hence the death rate $\delta = \delta(a)$ is a function of age only; for other cases see [110]. Specifically, we take

$$\beta = \begin{cases} nb(a, p(t)) & \text{for } a_j \leq a \leq a_M < +\infty, \\ 0 & \text{for } 0 \leq a < a_j, \end{cases}$$

$$p(t) = \xi p_J(t) + (1 - \xi)\, p_A(t), \quad 0 \leq \xi \leq 1,$$

in the McKendrick model (2.6) where b satisfies the normalization (2.15) and n is the inherent net reproductive number. We are interested in intraspecific competition, so it is assumed that $\partial_p b < 0$. The model parameter ξ measures the relative effect that the populations of adults and juveniles have on the adult birth rates. An increase in ξ means greater intraclass competition between juveniles and adults.

Theorem 2.2.1 implies that there exists an unbounded continuum of positive equilibria that bifurcates from the extinction equilibrium $\rho = 0$ at $n = 1$. From formula (2.21)

$$0 < n_1 = -\pi_0 \int_{a_j}^{a_M} \Pi(a) b(a) \partial_p b(a, 0) da,$$

where

$$\pi_0 = \xi \int_0^{a_j} \omega_J(a) \Pi(a) da + (1 - \xi) \int_{a_j}^{a_M} \omega_A(a) \Pi(a) da,$$

and hence the bifurcation is to the right and stable. In fact, $\partial_p b < 0$ and $\Pi(a) \leq 1$ imply that $\nu(p) < 1$ in (2.19) and it follows that $n > 1$ for all positive equilibria.

An interesting question concerns the "stabilizing" or "destabilizing" effects of juvenile versus adult competition. In a seminal work, Ebenman concluded from a simple (semelparous) discrete matrix model that such competition is "destabilizing" [151], [152], [154] (but see [299]). Discussion of this question requires, of course, a careful discussion of what is meant by "stabilizing" or "destabilizing" effects. For this purpose Ebenman uses the size of the parameter region in which stable positive equilibria occur and treats a shrinkage of this region as "destabilizing." Other measures of destabilization can be used. For example, the effect on equilibrium of total population size or the "resilience" of the equilibrium (the magnitude of the eigenvalue with smallest absolute value) [110]. A decrease in equilibrium level or resilience can also be considered destabilizing.

In the juvenile/adult model above we find that

$$\rho_e(a) = \frac{\nu^{-1}(1/n)}{\pi_0} \Pi(a)$$

for those $n > 1$ such that $1/n$ lie in the range of $\nu(p) = \int_{a_j}^{a_M} b(a,p)\Pi(a)da$, i.e., for $1 < n < (\lim_{p\to+\infty} \nu(p))^{-1}$. The sign of the derivative

$$\frac{\partial}{\partial \xi}\rho_e(a) = \left(\int_{a_j}^{a_M} \omega_A(a)\Pi(a)da - \int_0^{a_j} \omega_J(a)\Pi(a)da \right) \frac{\nu^{-1}(1/n)}{\pi_0^2}\Pi(a)$$

can be positive or negative. For example, for all age classes a

$$\begin{aligned} \frac{\partial}{\partial \xi}\rho_e(a) &< 0 \quad \text{for } a_j \approx a_M, \\ \frac{\partial}{\partial \xi}\rho_e(a) &> 0 \quad \text{for } a_j \approx 0. \end{aligned}$$

That is to say, if the maturation period a_j is long, then an increase in the interclass competitive effects results in decreased equilibrium levels for all age classes. On the other hand, if the maturation period a_j is short, then such an increase has the opposite effect of increasing equilibrium levels. We conclude for this case that interclass juvenile and adult competition is deleterious if the juvenile period is long but is advantageous if it is short. On the other hand, it is shown in [110] that equilibrium resilience for $n > 1$ near 1 is always decreased by an increase in the interclass competition.

Furthermore, it is interesting to note that

$$\frac{\partial}{\partial \xi}\left(\frac{\rho_e(a)}{\int_0^{a_M} \rho_e(a)da} \right) = 0$$

and hence that the normalized age distribution is unaffected by a change in the intraclass competition between juveniles and adults.

From the example above we see that the effects of juvenile/adult competition can be varied. The consequences of such a competition can be even more complicated if, unlike in the case considered above, the death rate δ is also affected by the competition. When δ also depends on a weighted total population size, say $\delta = \delta(a,p)$, $\partial_p(a,p) < 0$, where $p = (1-\varepsilon)p_J(t) + \varepsilon p_A(t)$, $0 \le \varepsilon \le 1$, Theorems 2.2.1 and 2.2.2 still imply the bifurcation of an unbounded continuum of positive equilibria which is to the right and stable. Equilibrium levels and resiliency in this case are studied in [110] where either deleterious or advantageous effects of increased interclass juvenile/adult competition can occur depending on the detailed age-specific nature of the competition. For example, if the population has two distinct reproduction ages which are sufficiently far apart, then juvenile/adult competition leads to increased equilibrium resilience [110].

Allee effects. Consider a McKendrick model (2.6) in which there is an Allee effect in adult fertility and death rates as functions of a single weighted total population size

$$p(t) = \int_0^{a_M} \omega(a)\rho(t,a)da.$$

That is to say, $\delta = \delta(a,p) \geq 0$, where

$$\begin{aligned} \partial_p \delta(a,p) &\leq 0 \quad \text{for } 0 < p < p_{cr}, \\ \partial_p \delta(a,p) &\geq 0 \quad \text{for } p_{cr} < p \end{aligned}$$

and $\beta = nb(a,p) \geq 0$, $\int_0^{a_M} b(a)\Pi(a)da = 1$, where

$$\begin{aligned} \partial_p b(a,p) &> 0 \quad \text{for } p \in (0, p_{cr}), \\ \partial_p b(a,p) &< 0 \quad \text{for } p \in (p_{cr}, +\infty), \\ \lim_{p\to+\infty} b(a,p) &= 0 \qquad \text{uniformly for } a \in [0, a_M]. \end{aligned}$$

These assumptions describe an Allee effect at low population densities $p < p_{cr}$ but a negative feedback effect at high densities $p > p_{cr}$. Note that $\partial_p b(a, p_{cr}) = \partial_p \delta(a, p_{cr}) = 0$ and $\partial_p^2 \delta(a, p_{cr}) \geq 0$, $\partial_p^2 b(a, p_{cr}) \leq 0$. For technical reasons we assume that

$$\partial_p^2 b(a, p_{cr}) < 0.$$

From (2.21), $n_1 < 0$ and the bifurcation in Theorems 2.2.1 and 2.2.2 is to the left and unstable.

By (2.19) we have

$$n\nu(p) = 1, \qquad \nu(p) \doteq \int_0^{a_M} b(a,p)\Pi(a,p)da,$$

where, from the assumptions made,

$$\text{(2.25)} \qquad \begin{aligned} \nu'(p) &> 0 \quad \text{for } p \in (0, p_{cr}), \\ \nu'(p) &< 0 \quad \text{for } p \in (p_{cr}, +\infty), \\ \nu''(p_{cr}) &< 0, \quad \lim_{p\to+\infty} \nu(p) = 0. \end{aligned}$$

Note that for a given pair $(n,p) \in R_+^2$ satisfying $n\nu(p) = 1$ there exists a unique positive equilibrium

$$\rho(a) = \frac{p}{\int_0^{a_M} \omega(a)\Pi(a,p)da}\Pi(a,p).$$

Define

$$n_{cr} \doteq \frac{1}{\nu(p_{cr})} \in (0,1).$$

From $n\nu(p) = 1$ and (2.25) we deduce the following facts (see Fig. 2.3).

(1) The extinction equilibrium is (locally asymptotically) stable for $n \in (0,1)$ and unstable for $n \in (1, +\infty)$.

(2) For $n \in (0, n_{cr})$ there exists no positive equilibrium.

(3) For $n \in (n_{cr}, 1)$ there exist two positive equilibria $\rho_1(a,n)$ and $\rho_2(a,n)$. Let $p_i(n) = \int_0^{a_M} \omega(a)\rho_i(a,n)da$. Then $0 < p_1(n) < p_2(n)$ and $\nu'(p_1(n)) > 0$ and

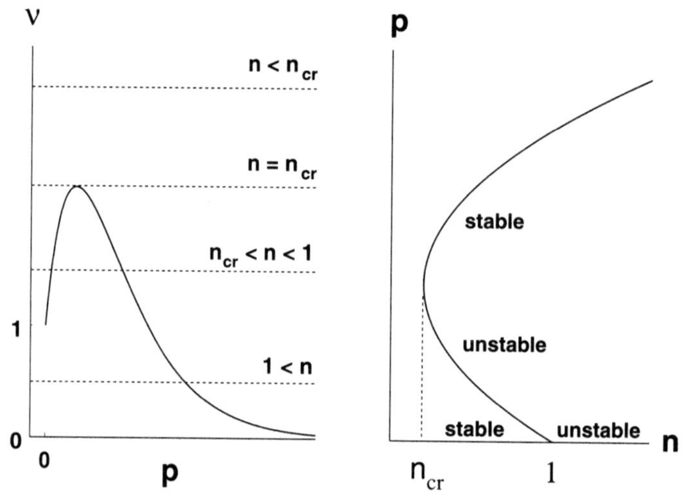

FIG. 2.3. *Equilibria are given by the intersection of the graph of $\nu(p)$ with a horizontal line located at $1/n$.*

$\nu'(p_2(n)) < 0$. By Theorem 2.2.2 the "lower" equilibrium $\rho_1(a,n)$ is unstable for $n \approx 1$. Also, $\lim_{n\to 1} p_1(n) = 0$ and $\lim_{n\to n_{cr}} p_1(n) = \lim_{n\to n_{cr}} p_2(n) = p_{cr}$.

(4) For $n \in (1, +\infty)$ there exists only one positive equilibrium $\rho_2(a,n)$ and $\nu'(p_2(n)) < 0$.

In specific models, it is usually the case that the "lower" equilibrium $\rho_1(t,n)$ is unstable for *all* $n \in (n_{cr}, +\infty)$ and the "upper" equilibrium, which meets the lower equilibrium at $n = n_{cr}$, is (locally asymptotically) stable at least for $n \approx n_{cr}$. That is to say, usually a saddle-node bifurcation occurs at the "turn around" point $n = n_{cr}$.

As an example, consider the case $\delta = \delta(a)$ when the death rate is not density dependent. The linearization of the McKendrick model at an equilibrium

$$\rho(a) = \frac{p}{\int_0^{a_M} \omega(a)\Pi(a)da}\Pi(a)$$

yields the equations (2.22), namely,

$$\begin{aligned} y'(a) &= -(\zeta + \delta(a))\, y(a), \\ y(0) &= n\int_0^{a_M} b(a,p)y(a)da + n\int_0^{a_M} \partial_p b(a,p)\rho(a)da \int_0^{a_M} \omega(\alpha)y(\alpha)d\alpha \end{aligned}$$

for complex ζ and nontrivial $y(a)$. The solution from the first equation $y(a) = y(0)e^{-\zeta a}\prod(a)$, $y(0) \neq 0$, when substituted into the second equation, yields the characteristic equation $c(\zeta, p) = 1$ for ζ, where

$$\begin{aligned} c(\zeta, p) &\doteq n \int_0^{a_M} \left[b(a,p) + \omega(a) \int_0^{a_M} \partial_p b(\alpha, p)\rho(\alpha)d\alpha \right] \Pi(a)e^{-\zeta a}da \\ &= n \int_0^{a_M} b(a,p)\Pi(a)e^{-\zeta a}da \\ &\quad + n\frac{p}{\int_0^{a_M}\omega(a)\Pi(a)da}\nu'(p) \int_0^{a_M} \omega(a)\Pi(a)e^{-\zeta a}da. \end{aligned}$$

The equilibrium is (locally asymptotically) stable if there exist no roots ζ satisfying $\operatorname{Re}\zeta \geq 0$ and is unstable if there exists at least one root satisfying $\operatorname{Re}\zeta > 0$. Note that

$$\lim_{\zeta=\text{real}\to+\infty} c(\zeta, p) = 0, \quad c(0,p) = 1 + n\nu'(p)p.$$

As noted above, $\nu'(p_1) > 0$ for the lower equilibrium $\rho = \rho_1$ with $n \in (n_{cr}, 1)$. Hence $c(0, p_1) > 1$ and it follows that there exists a positive real root $\zeta > 0$ of $c(\zeta, p_1) = 1$. Thus, *the lower equilibrium* ρ_1 *is unstable for all* $n \in (n_{cr}, 1)$.

To illustrate the saddle-node bifurcation at $n = n_{cr}$ we consider a specific example. Take $a_M = +\infty$, $\delta =$ constant > 0, and $\omega(a) \equiv 1$. Take $b(p) = \delta(1 + \frac{1}{c}p)[1-p]_+$, $0 < c < 1$, where $[x]_+ = x$ if $x \geq 0$ and $[x]_+ = 0$ if $x < 0$. Then $\nu(p) = (1 + \frac{1}{c}p)[1-p]_+$ and (2.19) becomes $n(1 + \frac{1}{c}p)[1-p]_+ = 1$. For $n_{cr} < n < 1$, $n_{cr} \doteq 4c(c+1)^{-2} < 1$, the solution of this quadratic yields two positive roots, p_1 and p_2, that satisfy $p_1 < p_{cr} < p_2 < 1$, where $p_{cr} \doteq (1-c)/2 < 1$. At $n = n_{cr}$ both equilibria equal p_{cr}. A calculation yields

$$c(\zeta, p_i) = \delta \frac{c + 2(1-c)p_i - 3p_i^2}{(c+p_i)(1-p_i)(\delta+\zeta)}$$

and the roots $\zeta = 2\delta p_i (p_{cr} - p_i)(c+p_i)^{-1}(1-p_i)^{-1}$ of the characteristic equations $c(\zeta, p_i) = 1$. Thus, $\zeta > 0$ for p_1 and this lower equilibrium is unstable and $\zeta < 0$ for p_2 and the upper equilibrium is stable.

2.4 Multispecies interactions

The interaction of several age-structured species can be modeled by McKendrick equations of the form (2.6) by allowing that a species' vital birth and death rates, β and δ, depend on the total population sizes of other species. Such models permit the description of age specific interactions among species, e.g., age specific differences in vulnerability to predation, in competitive efficiency for resources, etc. Such an assumption leads to a coupled system of McKendrick models. For example, a system of this kind is

$$\begin{aligned} &\partial_t \rho_i + \partial_a \rho_i = -\delta_i(a, p_{i1}, \ldots, p_{ik})\rho_i, \; t > 0, \; 0 < a < a_{M_i}, \\ &\rho_i(t,0) = \textstyle\int_0^{a_{M_i}} \beta_i(a, p_{i1}, \ldots, p_{ik})\rho_i(t,a)da, \; t > 0, \end{aligned} \tag{2.26}$$

where $\rho_i = \rho_i(t,a)$ is the age density function for the ith species, $i \in I[1,k]$, and

$$p_{ij}(t) \doteq \int_0^{a_{M_i}} \omega_{ij}(a)\rho_j(t,a)da$$

are weighted total population sizes.

One approach to the equilibrium existence and stability question is that taken in section 1.5 for multispecies interactions modeled by discrete matrix equations. Namely, the subcommunity of species $i \in I[2,k]$ is assumed to have a stable positive equilibrium $\rho_i^0(a)$ in the absence of the first species ρ_1, which is viewed as an "invader" species. The loss of stability of the subcommunity equilibrium $\rho_i^0(a)$ and the possible bifurcation of a positive equilibrium for the full community of species including ρ_1 can be studied as a function of a model parameter, say the inherent net reproductive number n of ρ_1 when the subcommunity is at its equilibrium.

Toward this end, we write

$$\beta_1 = nb_1(a, p_{11}, p_{12}, \dots, p_{1k}),$$

where b_1 is normalized by

$$\int_0^{a_{M1}} b_1\left(a, 0, p_{12}^0, \dots, p_{1k}^0\right) \Pi_1\left(a, 0, p_{12}^0, \dots, p_{1k}^0\right) da = 1,$$

$$\Pi_i(a, p_{11}, p_{12}, \dots, p_{1k}) \doteq \exp\left(-\int_0^a \delta_i(\alpha, p_{11}, p_{12}, \dots, p_{1k})d\alpha\right).$$

Bifurcation results follow immediately from theorems in [91], where even more general systems are considered. Or, alternatively, one can take advantage of the dependence of δ_i and β_i on the functionals $p_{ij}(t)$ in (2.26), which allows the equilibrium equations to be reduced to algebraic equations in the same way they were for the single species case in section 2.2, and independently derive bifurcation results using the theorems in Appendix B in a manner analogous to that carried out for multispecies matrix models in section 1.5. Either way, the end result is a continuum of positive (coexistence) equilibria, in which ρ_1 is present, that bifurcates at the critical value $n = 1$ from the subcommunity "trivial" equilibrium $\left(0, \rho_2^0, \dots, \rho_k^0\right)$, in which ρ_1 is absent. This continuum can be parameterized locally (by Liapunov/Schmidt expansions) and the direction of bifurcation determined, i.e., whether the positive coexistence equilibria near the bifurcation equilibrium exist for $n > 1$ or $n < 1$.

The equilibrium stability question is rigorously a difficult one for partial differential equations like (2.26); see [429] for a nonlinear semigroup treatment of stability for McKendrick equations. Formally, local (linearized) stability can be studied by investigating (time) exponential solutions of the equations linearized at the equilibrium of interest. This is done in [91] with the result that linearized stability of the positive equilibria can be related to the direction of bifurcation, at least in a neighborhood of the bifurcation point. Thus, it turns out (ignoring some technical details) that the subcommunity equilibrium loses stability as n is increased through $n = 1$ and the bifurcating coexistence (positive) equilibria are stable if the bifurcation is to the right and unstable if the bifurcation is to the left.

Consider the case of two interacting species ($k = 2$). Assume that the vital birth and death rates depend on only one weighted total population size of each species. Thus,

$$\begin{aligned} &\partial_t \rho_i + \partial_a \rho_i = -\delta_i\left(a, p_1, p_2\right) \rho_i, \; t > 0, \; 0 < a < a_{M_i}, \\ &\rho_i(t, 0) = \textstyle\int_0^{a_{M_i}} \beta_i\left(a, p_1, p_2\right) \rho_i(t, a) da, \; t > 0, \end{aligned} \tag{2.27}$$

where

$$p_j(t) = \int_0^{a_{M_i}} \omega_j(a) \rho_j(t, a) da$$

and

$$\beta_1(a, p_1, p_2) = n b_1(a, p_1, p_2),$$

$$\int_0^{a_{M1}} b_1(a, 0, p_2^0) \Pi_1(a, 0, p_2^0) da = 1. \tag{2.28}$$

From the equilibrium equations

$$\begin{aligned} \partial_a \rho_i(a) &= -\delta_i\left(a, p_1, p_2\right) \rho_i(a), \; 0 < a < a_{M_i}, \\ \rho_i(0) &= \int_0^{a_{M_i}} \beta_i\left(a, p_1, p_2\right) \rho_i(a) da \end{aligned}$$

follow $n_i(p_1, p_2) = 1$, where $n_i\left(p_1, p_2\right) \doteq \int_0^{a_{M_i}} b_i\left(a, p_1, p_2\right) \Pi_i\left(a, p_1, p_2\right) da$ or

$$\begin{aligned} &\text{(a)} \qquad n\nu_1\left(p_1, p_2\right) = 1, \\ &\text{(b)} \qquad n_2(p_1, p_2) = 1, \end{aligned} \tag{2.29}$$

where

$$\nu_1\left(p_1, p_2\right) \doteq \int_0^{a_{M_1}} b_1\left(a, p_1, p_2\right) \Pi_1\left(a, p_1, p_2\right) da.$$

These algebraic equations are identical to those that arose in the two species matrix model studied in section 1.5.2, namely, (1.61). Thus, if species ρ_2 has negative feedback density effects (at least near equilibrium) so that

$$\partial_{p_2} n_2(0, p_2^e) < 0,$$

then

$$\begin{aligned} \det M &> 0 \text{ implies a stable bifurcation at } n = 1, \\ \det M &< 0 \text{ implies an unstable bifurcation at } n = 1, \end{aligned}$$

where

$$M \doteq \begin{bmatrix} \partial_{p_1} n_1 & \partial_{p_2} n_1 \\ \partial_{p_1} n_2 & \partial_{p_2} n_2 \end{bmatrix}$$

and the partial derivatives are evaluated at $(p_1, p_2) = (0, p_2^e)$ and $n = 1$. (If $\partial_{p_2} n_2(0, p_2^e) > 0$, these inequalities are reversed.)

Thus, if $\det M > 0$, then the species ρ_1 can successfully invade the second species (at low density) if $n > 1$ but not if $n < 1$.

The remarks in section 1.5.2 concerning predator–prey interactions and competition interactions apply to this continuous age-structured model. For examples of continuous age-structured predator–prey (and host–parasite) see [108], [216], [217], [218], [231], [286]. An age-structured competition model is studied in [109].

2.5 Other structured models

The McKendrick age-structured model (2.5) is derived heuristically from the Leslie matrix model in section 2.1 by a limiting process. The derivation is made rigorous in [409]. The Usher matrix model is also subjected to a similar limiting process in [409] in order to derive a continuous size-structured model. Under the assumption that all newborns have the same size s_b, the result is the set of model equations

$$\begin{aligned} &\partial_t \rho + \partial_s (\gamma \rho) = -\delta \rho, \quad t > 0, \quad s_b < s < s_M \leq +\infty, \\ &\gamma \rho|_{s=s_b} = \int_0^{s_M} \beta \rho(t,s) ds, \quad t > 0. \end{aligned} \tag{2.30}$$

Here s is a measure of an individual's "body size" (length, surface area, volume, biomass, etc.) and $\rho = \rho(t,s)$ is a size-specific density function. In these equations $\delta = \delta(t,s)$ and $\beta = \beta(t,s)$ are size-specific death and birth rates. The coefficient $\gamma = \gamma(t,s)$ is the "growth" rate of an individual of size s at time t. An initial condition $\rho(0,s) = \rho_0(s)$ is added to these equations. If the maximal size $s_M < +\infty$, then it is natural to require the additional boundary condition $\rho(t,s) = 0$, $t \geq 0$, $s \geq s_M$. If any of the vital rates δ, β, or γ depend on the density ρ, then the model becomes nonlinear. This generalized McKendrick model is equivalent to the age-structured McKendrick model (2.5) if $\gamma \equiv 1$. It can also be derived from first principles as a continuity equation (balance law) [323]. If all newborns are not of the same size, then the boundary condition must be modified; see [323]. Models in which there is more than one structuring variable have a similar form. For example, if both age and size are used as structuring variables, then the partial differential equation for $\rho = \rho(t,s,a)$ becomes

$$\partial_t \rho + \partial_a \rho + \partial_s (\gamma \rho) = -\delta \rho.$$

For a treatment of age-size-structure equations see [404].

With regard to equilibria, the bifurcation theoretic approach used above for both discrete and continuous age-structured models has been applied in [409] to the autonomous and nonlinear size-structured model (2.30) with $s_M = +\infty$, in which $\delta = \delta(s, \rho)$, $\beta = \beta(s, \rho)$ and $\gamma = \gamma(s, \rho)$ depend on population density. A global continuum of positive equilibria is shown to bifurcate from the extinction equilibrium as a function of the inherent net reproductive number

$$n = \int_{s_b}^{+\infty} (b(\sigma, 0)/\gamma(\sigma, 0)) \exp\left(- \int_{s_b}^{s} (\delta(\sigma, 0)/\gamma(\sigma, 0))\, d\sigma \right) ds.$$

If, as a special case, the density effects are through a dependence on one or more weighted total population sizes $\delta = \delta(s, p_1, \ldots, p_k)$, $\beta = \beta(s, p_1, \ldots, p_k)$, and $\gamma = \gamma(s, p_1, \ldots, p_k)$, then the finite dimensional bifurcation theorem Theorem B.1.1 (see Appendix B.1) can once again be used on the equilibrium equations in a manner almost identical to that used on the McKendrick model (2.6) in section 2.2.2. For an application that utilizes these kinds of continuous size-structured models see section 3.2.

Finally, we point out that the bifurcation approach to equilibria for autonomous age-structured models taken above has been applied to periodic solutions of periodic age-structured models [90]. Periodic structured models are ones in which the birth and/or death rates are periodic functions of time, as might be appropriate for a population in a periodically fluctuating (e.g., seasonal) environment.

CHAPTER 3

Population Level Dynamics

One of the basic goals of structured population dynamics is to form a bridge from the level of individual organisms to the total population level. The modeling methodologies in Chapters 1 and 2 set up dynamical equations for a class distribution or a density function of individuals. Often these equations are not themselves utilized and studied, but instead are used as a starting point for the derivation of dynamical equations for a population level quantity (such as total population numbers, biomass, etc.). In this way the gap between the individual level and the population level is mathematically bridged. Moreover, the equations at the population level are sometimes more analytically tractable than those at the individual level, being of a "simpler" type (e.g., an ordinary rather than a partial differential equation) or of lower dimension. In this chapter we look at several classes of structured models for which it is possible to derive equations for total population level quantities. These types of models are specialized, of course, but several applications in which these types arise will be given.

3.1 Ergodicity and nonlinear models

For general linear matrix equations we found in section 1.1.2 that normalized distributions always equilibrate and that the dynamics of total population size could be determined by a simple linear scalar difference equation. We saw in section 2.2.1 that an analogous result holds for linear continuous age-structured models where, in this case, the dynamics of total population size can be determined from a scalar linear ordinary differential equation. In this section we will see how, under certain restrictive conditions, a similar jump from individual level dynamics to population level dynamics can be made for certain types of nonlinear models.

3.1.1 Discrete matrix models. In one of his seminal papers, P. H. Leslie considered a nonlinear modification of the linear age-structured matrix model that now bears his name [282]. This modification describes a specialized situation in which density effects on mortality are independent of age. Thus, each transition probability $t_{i,i-1}$ in the Leslie projection matrix

$$P = \begin{bmatrix} f_{11} & f_{12} & \cdots & f_{1,m-1} & f_{1m} \\ t_{21} & 0 & \cdots & 0 & 0 \\ 0 & t_{32} & \cdots & 0 & 0 \\ \vdots & \vdots & \ddots & \vdots & \vdots \\ 0 & 0 & \cdots & t_{m,m-1} & 0 \end{bmatrix} \tag{3.1}$$

is reduced by the same fraction $1/q(p)$, a fraction that depends on total population size $p = |x|$. Moreover, since the entries f_{ij} are the per capita number of *surviving* offspring (per unit time), each of these entries in the matrix must also be multiplied by the factor $1/q(p)$. This results in a nonlinear Leslie projection matrix

$$\frac{1}{q(p)} \begin{bmatrix} f_{11} & f_{12} & \cdots & f_{1,m-1} & f_{1m} \\ t_{21} & 0 & \cdots & 0 & 0 \\ 0 & t_{32} & \cdots & 0 & 0 \\ \vdots & \vdots & \ddots & \vdots & \vdots \\ 0 & 0 & \cdots & t_{m,m-1} & 0 \end{bmatrix}, \tag{3.2}$$

where $q(p) \geq 1$ for $p \geq 0$. If $q(0) = 1$, then the entries $t_{i,i-1}$ are *inherent* transition rates, i.e., the probabilities of surviving from one age class to the next when the effects of density are ignored (e.g., when the population density is very low). Similarly, the f_{1j} are the *inherent* fertilities, i.e., fertility rates at very low population densities.

Leslie studied the case $q(p) = 1 + cp$, $c > 0$. He noted that the resulting nonlinear matrix model has a stable age class distribution, just as in the Fundamental Theorem of Demography (see Theorem 1.1.2) for linear matrix models, but unlike linear models the total population size $p(t)$ always asymptotically equilibrates. We can see why this is true by considering the matrix equation

$$x(t+1) = h(x(t))Px(t), \quad 0 \leq x(0) \neq 0, \tag{3.3}$$

where P is a (constant) irreducible and primitive matrix and

$$h \in C^0\left(R_+^m, (0,1]\right), \quad h(0) = 1. \tag{3.4}$$

From

$$\frac{x(t+1)}{p(t+1)} = \frac{h(x(t))Px(t)}{|h(x(t))Px(t)|} = \frac{Px(t)/w(t)}{|Px(t)/w(t)|}$$

we see that the normalized class distribution

$$\varphi(t) \doteq \frac{x(t)}{p(t)}$$

satisfies the recursive formula

$$\varphi(t+1) = \frac{P\varphi(t)}{|P\varphi(t)|}, \quad \varphi(0) = \frac{x(0)}{|x(0)|}. \tag{3.5}$$

Similarly, it is shown that the normalized class distribution

$$\psi(t) \doteq \frac{y(t)}{|y(t)|}$$

of the solution of the *linear* matrix equation

$$y(t+1) = Py(t), \quad y(0) = x(0)$$

also satisfies (3.5). Since recursion formula (3.5) has a unique solution it follows that $\varphi(t) = \psi(t)$ and (see section 1.1.2)

$$\lim_{t\to+\infty} \frac{x(t)}{p(t)} = \frac{v}{|v|}, \tag{3.6}$$

where v is a positive eigenvalue associated with the dominant eigenvalue $r > 0$.

Moreover,

$$p(t+1) = h\left(\varphi(t)p(t)\right)|P\varphi(t)|\,p(t), \quad p(0) = |x(0)| > 0. \tag{3.7}$$

This is a nonautonomous scalar difference equation for the dynamics of total population size $p(t)$ which is coupled to the normalized class distribution $\varphi(t)$ (i.e., to the level of individuals). However, since $\varphi(t)$ stabilizes, it can be replaced by its limit (3.6) in equation (3.7). This yields the *limiting equation*

$$q(t+1) = rh\left(vq(t)\right)q(t), \quad q(0) = |x(0)| > 0 \tag{3.8}$$

for total population size [278]. This limiting equation is a scalar autonomous difference equation that is uncoupled from the dynamics of the normalized class distribution.

Although it is natural to conjecture that the asymptotic dynamics of the original equation (3.7) for total population size $p(t)$ and those of its limiting equation (3.8) for $q(t)$ are closely related, this mathematical question does not seem to have been thoroughly studied. When the limiting equation (3.8) has "tame" asymptotic dynamics, a rigorous connection to the dynamics of equation (3.7) is made in the following theorem [95] (also see [7]).

THEOREM 3.1.1. *Assume that $P \geq 0$ is irreducible and primitive and h satisfies* (3.4). *Suppose that the limiting equation* (3.8) *has at most a finite number of cycles in any compact interval, all of which are hyperbolic, and every forward bounded solution tends to a cycle as $t \to +\infty$. Then a forward bounded solution $p(t)$ of* (3.7) *tends to a cycle of its limiting equation* (3.8) *as $t \to +\infty$. Moreover, if the extinction equilibrium $q = 0$ of the limiting equation is unstable, then the solution $p(t)$ of* (3.7) *does not tend to* 0 *as $t \to +\infty$.*

The cycle to which the bounded solution of (3.7) tends asymptotically in Theorem 3.1.1 is not necessarily a stable cycle of the limiting equation. Nor is it necessarily true that a solution $p(t)$ of equation (3.7) that starts sufficiently near a stable cycle of the limiting equation (3.8) will tend asymptotically to that cycle. This is because the dependency of the nonautonomous equation (3.7) on t can cause significant initial differences between its solutions and those of its limiting equation (3.8). However, if the total population size $p(t)$ starts near a stable cycle of the limiting equation and the normalized class distribution $\varphi(t)$ starts near the stable class distribution $v/|v|$, then presumably such differences would not arise. This is the content of the following theorem [95].

THEOREM 3.1.2. *Suppose that $q(t)$ is an asymptotically stable cycle of the limiting equation* (3.8). *Under the assumptions of Theorem* 3.1.1 *there exists a $\delta > 0$ such that $|p(0) - q(0)| < \delta$ and $|\varphi(t) - v| < \delta$ imply that* $\lim_{t\to+\infty} |p(t) - q(t)| = 0$.

Thus, in general, to guarantee that the population asymptotically approaches a stable cycle of the limiting equation it is not sufficient that the total population size start near the cycle, but it is required in addition that the initial normalized class distribution start near the stable class distribution. Or, put another way, a small perturbation of total population from a stable cycle might be destabilizing if the disturbance significantly perturbs the class distribution from the stable class distribution.

In Theorems 3.1.1 and 3.1.2 an equilibrium is a cycle of period 1. For more on limiting equations see [77], [278].

If $h = h(p)$ is a function of a weighted total population size

$$p(t) \doteq \omega^{\tau} x(t) = \sum_{i=1}^{m} \omega_i x_i(t), \quad \omega_i \geq 0, \quad \sum_{i=1}^{m} \omega_i \neq 0,$$

then the limiting equation is

$$q(t+1) = rh\left(\omega^{\tau} v q(t)\right) q(t), \quad q(0) = |x(0)| > 0.$$

In particular, if $h = h(p)$ is a function of total population size ($\omega_i = 1$), then the limiting equation is

$$q(t+1) = rh\left(q(t)\right) q(t), \quad q(0) = |x(0)| > 0.$$

The age-structured matrix equation with $h(p) = 1 + cp$ considered by Leslie yields the Beverton–Holt limiting equation

$$q(t+1) = r\frac{1}{1 + cq(t)} q(t), \quad q(0) = |x(0)| > 0,$$

where r is the strictly dominant eigenvalue of the Leslie matrix (3.1). This equation has only equilibrium dynamics, namely, $\lim_{t\to+\infty} q(t) = 0$ if $r < 1$ and $\lim_{t\to+\infty} q(t) = (r-1)/c$ if $r > 1$. Theorem 3.1.1 implies that $p(t)$ satisfies these same alternatives and, in either case, the normalized class distribution

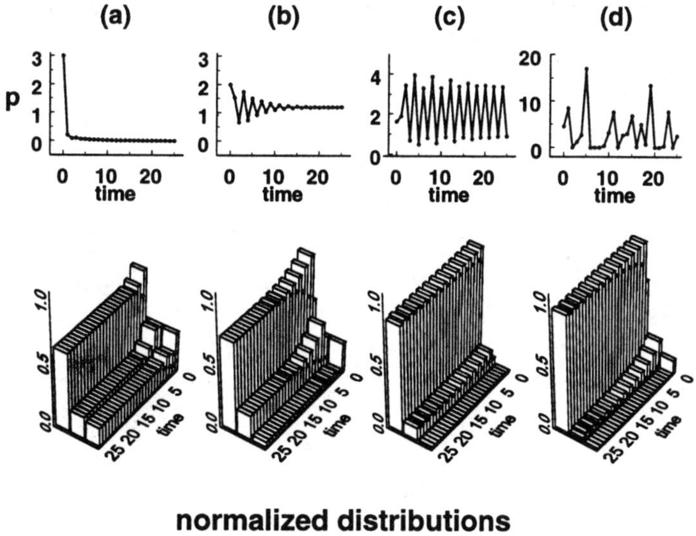

FIG. 3.1. *The normalized distributions* $\varphi(t)$ *from a* 3×3 *Leslie model* (3.2) *with a nonlinear factor* $1/q(p) = \exp(-p)$ *stabilize to* $v/|v|$ *while the total population size* $p(t)$ *exhibits several different asymptotic dynamics. In* (a) $f_{11} = 0$, $f_{12} = 2.25$, $f_{13} = 1$, $p_{21} = 0.25$, $p_{32} = 0.75$ *with dominant eigenvalue* $r = 0.88$ *and* $v/|v| = (0.65, 0.19, 0.16)$ *and the population goes to extinction. In* (b) $f_{11} = 0$, $f_{12} = 10$, $f_{13} = 10$, $p_{21} = 0.9$, $p_{32} = 0.75$ *with* $r = 3.3$ *and* $v/|v| = (0.75, 0.20, 0.05)$ *and the total population size tends to a positive equilibrium. In* (c) $f_{11} = 0$, $f_{12} = 75$, $f_{13} = 50$, $p_{21} = 0.9$, $p_{32} = 0.75$ *with* $r = 8.5$ *and* $v/|v| = (0.896, 0.095, 0.009)$ *and the total population tends to a cycle of period* 2. *In* (c) $f_{11} = 2$, $f_{12} = 750$, $f_{13} = 1$, $p_{21} = .9$, $p_{32} = 0.75$ *with* $r = 27.0$ *and* $v/|v| = (0.9669, 0.0322, 0.0009)$ *and the total population size appears chaotic.*

stabilizes to the unit eigenvector $v/|v| > 0$ of the Leslie matrix (3.1) associated with the dominant eigenvalue.

If in Leslie's nonlinear model $h(p) = \exp(-cp)$ is used instead, the limiting equation is the Ricker equation

$$q(t+1) = rq(t)\exp\left(-cp(t)\right), \quad q(0) = |x(0)| > 0$$

for which stable attractors other than equilibria are possible. For $r < 1$, $\lim_{t\to+\infty} q(t) = 0$, and for $1 < r < e^2$ there is a unique positive equilibrium $c^{-1}\ln r$ which is (globally) asymptotically stable. Theorem 3.1.1 implies that the same alternatives hold for the total population size $p(t)$. For $r > e^2$ there occurs the familiar period doubling cascade of stable cycles up to the critical value $r_{cr} \approx e^{2.6924} \approx 14.767$ (see [313]). Theorems 3.1.1 and 3.1.2 are applicable for $r < r_{cr}$. For $r > r_{cr}$ chaotic dynamics occur for the limiting equation and although in this case we have no theorems relating such dynamics to the dynamics of $p(t)$, the numerical simulations in Fig. 3.1 suggest that $p(t)$ can also be "chaotic."

A stable class distribution result can be obtained for more general equations than (3.3) from the following theorem (a proof of which appears in Appendix C) [98].

THEOREM 3.1.3. *Consider the equation*

$$x(t+1) = [a(t)I + b(t)L]\, x(t), \tag{3.9}$$

where

(a) $0 \le a(t) \le a_0$, $0 < b_0 \le b(t)$ *for constants* a_0 *and* b_0 *and* $t \in I[0, +\infty)$;

(b) I *is the* $m \times m$ *identity and* L *is an* $m \times m$ *matrix that has a strictly dominant, simple eigenvalue* $r > 0$ *with a positive eigenvector* $v > 0$.

Suppose that $x(t)$ *is a solution satisfying* $0 \ne x(t) \ge 0$ *for all* $t \in I[0, +\infty)$ *and* $p(t) \doteq \omega^\tau x(t) = \sum_{i=1}^m \omega_i x_i(t)$, $0 \le \omega_i$, $\sum_{i=1}^m \omega_i \ne 0$, *is a weighted total population size. Then*

$$\lim_{t\to+\infty} \frac{x(t)}{p(t)} = \frac{v}{\omega^\tau v}. \tag{3.10}$$

This theorem can be applied when a and b are functions of x and equation (3.9) is nonlinear. Using (3.10) we can derive a limiting equation for the weighted total population size $p(t)$ as follows. Substituting $x(t) = \varphi(t)p(t)$ into

$$x(t+1) = [a(x(t))I + b(x(t))L]\, x(t)$$

and replacing $\varphi(t)$ by its limit $v/\omega^\tau v$ in the result, we obtain (relabeling p as q)

$$\frac{v}{\omega^\tau v} q(t+1) = \left[a\left(\frac{v}{\omega^\tau v} q(t)\right) + rb\left(\frac{v}{\omega^\tau v} q(t)\right)\right] \frac{v}{\omega^\tau v} q(t)$$

and hence the scalar limiting equation

$$q(t+1) = \left[a\left(\frac{v}{\omega^\tau v} q(t)\right) + rb\left(\frac{v}{\omega^\tau v} q(t)\right)\right] q(t).$$

If $a = a(p)$ and $b = b(p)$ are functions of p, then the limiting equation simplifies to the scalar equation

$$q(t+1) = [a(q(t)) + rb(q(t))]\, q(t). \tag{3.11}$$

In this theorem the dependence of the coefficients a and b on t can be implicit through a dependence on x. Thus, equation (3.3) is included in (3.9) by setting $a(t) \equiv 0$, $b(t) \equiv h(x(t))$, and $L = P$.

Theorem 3.1.3 was motivated by size-structured models of the following type. In a matrix equation with projection matrix $P = T + F$ write the transition rates t_{ij} in $T = [t_{ij}]$ as

$$t_{jj} = \pi_j (1 - \gamma_j), \quad t_{ij} = \pi_i \tau_{ij} \gamma_i,$$

where γ_j is the fraction that leaves the jth class per unit time, τ_{ij} is the fraction of those who leave that move to class i, and π_j is the survival rates per unit time. Make the following two simplifying assumptions: γ_j and the class specific fertility rates f_{ij} in the fertility matrix F are proportional to a common resource consumption rate $u \geq 0$ and the survival rates π_j are class independent. Thus,

$$\gamma_j = \tau_j u, \quad f_{ij} = \varphi_{ij} u, \quad \pi_i = \pi,$$

where it is required that $\tau_j u \leq 1$. In this way assumptions about how individuals of each structuring class consume a designated resource and how that resource consumption affects both fertility and the movement between classes (e.g., growth to other size classes or changes to other life-cycle stages) can be included in the model by construction of a submodel for u and its dependence on resource density, competition factors, etc.

Selecting any index k we can write the projection matrix in the form $a(t)I + b(t)L$, where

$$a = \pi\left(1 - \tau_k u\right), \quad b = \pi u,$$

and L is the matrix

$$\begin{bmatrix} \tau_k - \tau_1 + \varphi_{11} & \tau_{12}\tau_2 + \varphi_{12} & \cdots & \tau_{1k}\tau_k + \varphi_{1k} & \cdots & \tau_{1m}\tau_m + \varphi_{1m} \\ \tau_{21}\tau_1 + \varphi_{21} & \tau_k - \tau_2 + \varphi_{22} & \cdots & \tau_{2k}\tau_k + \varphi_{2k} & \cdots & \tau_{2m}\tau_m + \varphi_{2m} \\ \vdots & \vdots & \ddots & \vdots & \ddots & \vdots \\ \tau_{k1}\tau_1 + \varphi_{k1} & \tau_{k2}\tau_2 + \varphi_{k2} & \cdots & \varphi_{kk} & \cdots & \tau_{km}\tau_m + \varphi_{km} \\ \vdots & \vdots & \ddots & \vdots & \ddots & \vdots \\ \tau_{m1}\tau_1 + \varphi_{m1} & \tau_{m2}\tau_2 + \varphi_{m2} & \cdots & \tau_{mk}\tau_k + \varphi_{mk} & \cdots & \tau_k - \tau_m + \varphi_{mm} \end{bmatrix}.$$

If k chosen as the index for which τ_k is maximal ($\tau_k \geq \tau_i$ for all i), then $L \geq 0$. If the final simplifying assumption is made that the only model parameter affected by population density is the resource uptake function $u = u(x)$, where $u(0) = 1$, then the model equations take the form (3.9) with $a(t) \doteq \pi\left(1 - \tau_k u(x(t))\right)$, $b(t) \doteq \pi u(x(t))$, i.e.,

$$x(t+1) = \pi\left[\left(1 - \tau_k u(x(t))\right) I + u(x(t)) L\right] x(t). \tag{3.12}$$

If L satisfies the requirement in Theorem 3.1.3 (which it will, for example, if it is nonnegative, irreducible, and primitive) and if, for a given solution $x(t)$, the coefficients a and b satisfy the requirements of this theorem, then $x(t)$ will have a stable normalized class distribution (3.10). The requirement on a is satisfied with $a_0 = 1$. The condition on b requires, for each solution $x(t) \geq 0$, that the uptake function $u(x(t))$ be bounded away from 0 for all t. This condition will be met, for example, if $u(x) > 0$ is continuous for all $x \geq 0$ and the solution is bounded.

The limiting equation is

$$q(t+1) = \pi\left[1 + \theta \upsilon(q(t))\right] q(t),$$

where $\upsilon(q) \doteq u(vq/\omega^{\tau} v)$, $\upsilon(0) = 1$, and $\theta \doteq r - \tau_k$. While this is the limiting equation for the population level quantity $p(t)$, individual (class) level parameters contribute to the dynamics through the eigenvalue r and eigenvector v

of the matrix L. Note that if $u = u(p)$ is a function of the weighted total population size p, then $v(q) = u(q)$ in the limiting equation.

As an example consider the self regulatory case when the resource uptake function $u(x) \geq 0$, $u(0) = 1$, is a decreasing function of each class density x_i. Then $v(q)$ is a decreasing function of the real variable $q > 0$ and consequently $v(q) < 1$ for $q > 0$. From the inequality $q(t+1) \leq \pi(1+\theta)q(t)$ we find $\lim_{t\to+\infty} q(t) = 0$ for all $q(0) \geq 0$, and the population goes extinct if $\theta < \theta^{cr} \doteq (1-\pi)/\pi$ (Theorem 3.1.1). If $\theta > \theta_{cr}$, then there exists a unique positive equilibrium of the limiting equation $q = v^{-1}(\theta^{cr}/\theta)$. This equilibrium is asymptotically stable near the bifurcation point $\theta \approx \theta^{cr}$, but might lose stability for larger θ (for which the familiar period doubling route to chaos might occur). For example, for the case $u(p) = 1 + cp$ studied by Leslie the positive equilibrium is globally asymptotically stable for all $\theta > \theta^{cr}$, while for the Ricker nonlinearity $u(p) = \exp(-cp)$ a period doubling route to chaos occurs in the limiting equation as θ is increased.

For other generalizations of Theorem 3.1.3 and applications see [78], [79], [289]. For more on ergodic theorems see [7], [64], [65], [66], [67], [177], [186].

A size-structured model. A size-structured model of the form (3.12) is studied in [98], [101]. In that model individuals are classified according to body length and the projection matrix is a nonlinear Usher matrix of the form

$$\pi \begin{bmatrix} 1-\beta_1 e^{-dp}+\alpha_1 e^{-dp} & \alpha_2 e^{-dp} & \cdots & \alpha_m e^{-dp} \\ \beta_1 e^{-dp} & 1-\beta_2 e^{-dp} & \cdots & 0 \\ 0 & \beta_2 e^{-dp} & \cdots & 0 \\ \vdots & \vdots & \ddots & \vdots \\ 0 & 0 & \cdots & 0 \\ 0 & 0 & \cdots & 1 \end{bmatrix}. \tag{3.13}$$

All newborns lie in the smallest (first) size class and an individual either remains in its size class or grows to the next in one unit of time. No individual grows larger than the final size class m.

The parameters α_i and β_i are called the *reproductive efficiency* and *growth efficiency* coefficients, respectively. They are related, in this model, to various size specific parameters, namely,

$$\alpha_j = \frac{(1-\kappa_j)r_j\sigma_j s_j^2}{\omega_j w_1}, \qquad \beta_j = \frac{\kappa_j r_j \sigma_j}{3\mu_j\eta_j\delta_j} < 1, \tag{3.14}$$

where r_j is the inherent (low density) resource uptake rate (per unit time per unit body surface area) of an individual from the jth size class; e^{-d} denotes the fractional decrease in resource uptake of any individual (per unit time per unit body surface area) due to a unit surface area of a competitor; s_j is the body length of an individual from the jth size class (a representative size from the size interval, say the middle point); δ_j is the size class interval length; σ_j is a proportionality factor that relates body surface area to the square of body length

s_j^2 (and hence depends on the geometry of an individual); μ_j is a body density per unit volume (so that $\mu_j s_j^3$ is body weight); κ_j is the fraction of consumed resource that individuals of size s_j allocate to growth, and $1-\kappa_j$ is the fraction of consumed resource that individuals of size s_j allocate to reproduction; η_j is a conversion factor of resource units to body weight; ω_j is a conversion factor of resource units to offspring body weight; and finally $w_1 = \mu_1 s_1^3$ is the weight at birth. In this model p is the weighted total population size $p = \sum_{i=1}^{m} \sigma_i s_i^2 x_i(t)$; i.e., p is the total population surface area.

These details are given in order to show the considerable amount of class specific (individual level) information that is included within the strictly dominant eigenvalue r of the matrix

$$L = \begin{bmatrix} \beta_k - \beta_1 + \alpha_1 & \alpha_2 & \cdots & \alpha_k & \cdots & \alpha_{m-1} & \alpha_m \\ \beta_1 & \beta_k - \beta_2 & \cdots & 0 & \cdots & 0 & 0 \\ \vdots & \vdots & \ddots & \vdots & \ddots & \vdots & \vdots \\ 0 & 0 & \cdots & 0 & \cdots & 0 & 0 \\ \vdots & \vdots & \ddots & \vdots & \ddots & \vdots & \vdots \\ 0 & 0 & \cdots & 0 & \cdots & \beta_k - \beta_{m-1} & 0 \\ 0 & 0 & \cdots & 0 & \cdots & \beta_{m-1} & \beta_k \end{bmatrix} \tag{3.15}$$

and hence within the single parameter $\theta = r - \beta_k$ that determines the dynamical properties of the limiting equation

$$q(t+1) = \pi \left[1 + \theta e^{-dq(t)}\right] q(t)$$

for total population surface area. In this way the effect that a particular size specific parameter has on the population level dynamics can be determined by studying the effect that the parameter has on θ.

As a function of θ, solutions of the limiting equation tend to 0 if $\theta < \theta_{cr}$. For $\theta > \theta_{cr}$ solutions persist, but there is a period doubling cascade to chaos as θ is increased. Note that θ is the eigenvalue of

$$L - \beta_k I = \begin{bmatrix} -\beta_1 + \alpha_1 & \alpha_2 & \cdots & \alpha_k & \cdots & \alpha_{m-1} & \alpha_m \\ \beta_1 & -\beta_2 & \cdots & 0 & \cdots & 0 & 0 \\ \vdots & \vdots & \ddots & \vdots & \ddots & \vdots & \vdots \\ 0 & 0 & \cdots & 0 & \cdots & 0 & 0 \\ \vdots & \vdots & \ddots & \vdots & \ddots & \vdots & \vdots \\ 0 & 0 & \cdots & 0 & \cdots & -\beta_{m-1} & 0 \\ 0 & 0 & \cdots & 0 & \cdots & \beta_{m-1} & 0 \end{bmatrix}$$

with largest real part.

Perhaps the simplest example is the case of $m = 2$ size classes in which the smaller consists of juveniles ($\kappa_1 = 1$ and $\alpha_1 = 0$) and the larger consists of adults ($\kappa_2 < 1$ and $\alpha_2 > 0$). Then the projection matrix is

$$\pi \begin{bmatrix} 1 - \beta e^{-dp} & \alpha e^{-dp} \\ \beta e^{-dp} & 1 \end{bmatrix} \tag{3.16}$$

(in which unnecessary subscripts have been dropped: $\beta_1 = \beta$, $\alpha_2 = \alpha$ and $\kappa_2 = \kappa$) and

$$L = \begin{bmatrix} 0 & \alpha \\ \beta & \beta \end{bmatrix}, \qquad L - \beta I = \begin{bmatrix} -\beta & \alpha \\ \beta & 0 \end{bmatrix}$$

from which we calculate

$$\theta = \frac{-\beta + \left(\beta^2 + 4\alpha\beta\right)^{1/2}}{2}. \tag{3.17}$$

As seen above the population goes extinct if $\theta < \theta^{cr} = (1-\pi)/\pi$ and persists if $\theta > \theta^{cr}$, so it is advantageous for the population to increase θ. Since θ is an increasing function of each coefficient α and β, it is to the population's advantage to increase either coefficient. How changes in the size class specific parameters listed above cause increases or decreases in these reproductive or growth coefficients can be determined from the formulas (3.14), i.e.,

$$\alpha = \frac{(1-\kappa)\, r\sigma s^2}{\omega w}, \qquad \beta = \frac{\kappa r\sigma}{3\mu\eta\delta} < 1$$

(where once again unnecessary subscripts have been dropped).

Consider, for example, θ as a function of the growth allocation fraction κ of adults. Writing

$$\theta(\kappa) = c_1\left(-\kappa + \left(\kappa^2 + c_2\,(1-\kappa)\,\kappa\right)^{1/2}\right), \tag{3.18}$$

where

$$c_1 \doteq \frac{r\sigma}{6\mu\eta\delta}, \qquad c_2 \doteq \frac{12\mu\eta\delta s^2}{\omega w}$$

are dimensionless parameters, it is straightforward to show that $\theta(\kappa)$ has a unique maximum value of $\theta_{\max} = c_1 c_2(\sqrt{c_2}+1)^{-1}/2$ which occurs for the unique fraction $\kappa = \kappa_{\max}(c_2) \doteq \frac{1}{2}(c_2 - \sqrt{c_2})\,(c_2-1)^{-1}$. It is interesting to note that $\kappa_{\max}$ lies between 0 and 0.5 and consequently the maximum value of θ is attained in this model by using (and only by using) an adult resource allocation that assigns more than half to reproduction. The fraction $\kappa_{\max}(c_2)$ is an increasing function of $c_2 > 0$ satisfying $\kappa_{\max}(0) = 0$, $\lim_{c_2 \to +\infty} \kappa_{\max}(c_2) = 0.5$. Thus, for c_2 large θ is maximized by nearly an equal resource allocation to growth and reproduction, while for c_2 small θ is maximized by allocating very little consumed resource to growth and nearly all to reproduction. See Fig. 3.2.

An interference competition model and nonequilibrium coexistence. In a coupled system of matrix equations for the dynamics of interacting structured species, scalar limiting equations can be derived for the total population sizes of those species whose matrix equation is of the type in Theorem 3.1.3. See [79] for examples. We consider a competition example in which all species

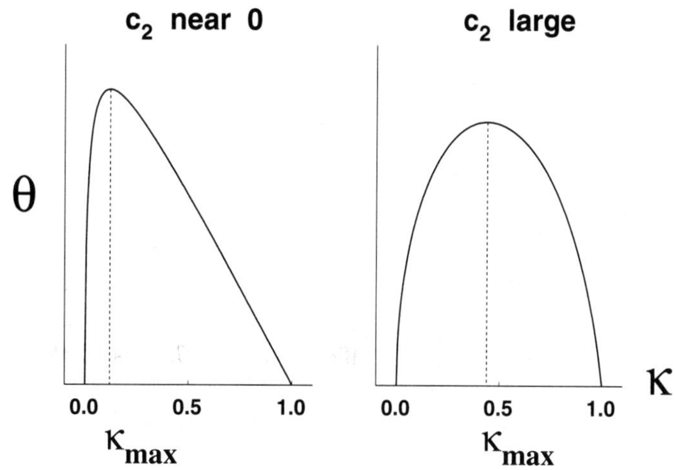

FIG. 3.2. *For small c_2 the maximum of θ is attained when only a small fraction κ_{max} of resource is allocated to adult growth. For large c_2 the maximum of θ is attained when the resource allocations to adult growth and reproduction are nearly equal.*

are described by matrix equations of this type and hence the system of matrix equations for species class distributions can be replaced by a limiting system of scalar equations.

Consider a community of n species, each of whose dynamics are modeled by a matrix equation of the type (3.13). Assume the species are in competition for a resource and that this competition is expressed by the dependence of a common uptake function u on the weighted population sizes $p_j(t)$ of all species. The weighted population sizes p_j do not necessarily have the same weights. Thus, we write $u = u(p_1, \dots, p_n)$, $u(0, \dots, 0) = 1$, and under the assumptions of Theorem 3.1.3 we obtain the system of limiting equations

$$q_i(t+1) = \pi_i \left[1 + \theta_i u(q_1(t), \dots, q_n(t))\right] q_i(t), \tag{3.19}$$

where π_i is now the class independent survival rate for the ith species and θ_i is computed from the L matrix for species i. How the dynamical properties of this population level competition model depend on class specific (individual level) parameters can be studied through the formulas (3.14) that relate how θ_i depends on the entries in the matrix (3.15) for the ith species.

Since we are interested in competition, we assume that $u(p_1, \dots, p_n) > 0$ is a decreasing function of each of its arguments

$$\frac{\partial u}{\partial p_i} < 0, \qquad i \in I[1, n]. \tag{3.20}$$

It follows that

$$0 < u(p_1, \dots, p_n) \leq 1, \qquad p_i \in R^1_+.$$

Define

$$\theta_i^{cr} \doteq \frac{1-\pi_i}{\pi_i}.$$

The first observation to make is that, arguing exactly as above for a single species, any species for which $\theta_i < \theta_i^{cr}$ will asymptotically go extinct. Therefore, we might as well assume that

$$\theta_i > \theta_i^{cr} \quad \text{for all} \quad i. \tag{3.21}$$

In this case, the limiting equation for each species has a positive equilibrium $q_i = q_i^e > 0$ given by the unique solution of the equation

$$1 = \pi_i \left(1 + \theta_i u(0, \dots, q_i^e, \dots, 0)\right).$$

The inequality (3.21) and the monotonicity assumption (3.20) imply that this equation has a unique positive equilibrium.

The second observation is that if a nonextinction equilibrium $(q_1, \dots, q_n) \geq 0$ exists, then the ratios θ_i^{cr}/θ_i would have to be equal for all subscripts corresponding to nonzero components $q_i > 0$. Ignoring this unlikely ("nongeneric") case it follows that the only positive nonextinction equilibria are those with a single positive component $q_i > 0$, which is in fact an equilibrium for the ith species in the absence of the other species, i.e., $q_i = q_i^e$. It is impossible, in this model, for two or more species to coexist *in a state of equilibrium*, a fact in keeping with the famous competitive exclusion principle in theoretical ecology.

The local stability of an equilibrium $(0, \dots, q_j^e, \dots, 0)$ is determined by the n eigenvalues

$$\pi_i \left(1 + \theta_i u(0, \dots, q_j^e, \dots, 0)\right) > 0 \quad \text{for all} \quad i \neq j \tag{3.22}$$

and

$$1 + \pi_j \theta_j q_j^e \frac{\partial u(0, \dots, q_j^e, \dots, 0)}{\partial p_j} \tag{3.23}$$

associated with the linearization of (3.19) at this equilibrium. Suppose that the jth species has a stable equilibrium in the absence of the other species; specifically assume that the absolute value of the eigenvalue (3.23) is less than 1. Stability is then determined by the remaining eigenvalues (3.22). Since

$$\pi_j \left(1 + \theta_j u(0, \dots, q_j^e, \dots, 0)\right) = 1$$

implies that

$$u(0, \dots, q_j^e, \dots, 0) = \theta_j^{cr}/\theta_j,$$

these eigenvalues can be written as

$$\pi_i \left(1 + \theta_i \frac{\theta_j^{cr}}{\theta_j}\right) \quad \text{for all } i \neq j.$$

These eigenvalues are less than 1 if and only if

$$\frac{\theta_i}{\theta_i^{cr}} < \frac{\theta_j}{\theta_j^{cr}}, \qquad i \neq j.$$

Thus, one and only one of the equilibria is (locally asymptotically) stable and it is the one corresponding to the largest ratio θ_j/θ_j^{cr}, say for $j = k$.

These results only suggest that all species except the kth species go extinct, since the stability analysis is only local. Under certain circumstances, however, the stability of the kth species equilibrium can be shown to be globally attracting. Namely, suppose

$$\pi_k > \pi_i \quad \text{and} \quad \theta_k \geq \theta_i \quad \text{for all} \qquad i \neq k.$$

Then θ_k/θ_k^{cr} is the largest ratio and $\delta \doteq \max_{i \neq k} \{\pi_i/\pi_k\} < 1$. Let $q_i(t)$, $i \in I[1,n]$, be a (forward) bounded solution of (3.19) with $q_i(0) > 0$. For $i \neq k$ we have from (3.19) that

$$0 < \frac{q_i(t+1)}{q_k(t+1)} \leq \delta \frac{q_i(t)}{q_k(t)}, \qquad t \in I[0, +\infty).$$

Consequently $q_i(t)/q_k(t) \to 0$ and $q_i(t) \to 0$ as $t \to +\infty$.

Through the relationship of the ratio θ_i/θ_i^{cr} to the size class specific parameters for the ith species in the model (σ_j, δ_j, η_j, etc.), obtained through the formulas (3.14) for the entries in the matrix (3.15), one can study how any one of these size class specific parameters affects the competitive outcome of the system.

As an example, consider the competition of two size-structured species, each structured by $m = 2$ size classes, modeled by a coupled system of two matrix equations with projection matrices of the form (3.16). Thus, the smallest size classes consist of (nonreproducing) juveniles. If the species are in competition for the resource, then $u = \exp(-d_1 p_1 - d_2 p_2)$, where p_i is the total population surface area of the ith species and the limiting equations of each species form the coupled system

$$\begin{aligned} q_1(t+1) &= \pi_1 \left[1 + \theta_1 \exp\left(-d_1 q_1(t) - d_2 q_2(t)\right)\right] q_1(t), \\ q_2(t+1) &= \pi_2 \left[1 + \theta_2 \exp\left(-d_1 q_1(t) - d_2 q_2(t)\right)\right] q_2(t). \end{aligned} \tag{3.24}$$

The surviving species is the one with the largest ratio $\theta_i/\theta_i^{cr} > 1$, where each θ_i is given by the formula (3.17) using the reproductive and growth coefficients for that species, i.e.,

$$\theta_i = \frac{-\beta_i + \left(\beta_i^2 + 4\alpha_i\beta_i\right)^{1/2}}{2}.$$

Consider two species that are identical except for their adult growth allocation fractions, which are different. By this we mean the two species have the same model parameter values (σ, δ, η, etc.) except for κ. Then both $\theta_i = \theta(\kappa_i)$ are given by the same function of κ (namely, (3.18)). It follows that if one species utilizes the allocation fraction $\kappa_{\max}$, then that species "wins" the competition.

The competition model (3.19) is consistent with the fundamental competitive exclusion principle from theoretical ecology insofar as equilibrium dynamics are concerned. However, from the broader perspective of nonequilibrium dynamics this famous principle does not always hold.

From a bifurcation theory point of view the competitive exclusion principle can be explained, for the type of competition models being considered here, as follows. Consider the two equations

$$1 = \pi_1 \left(1 + \theta_1 u(q_1, q_2)\right), \qquad 1 = \pi_2 \left(1 + \theta_2 u(q_1, q_2)\right)$$

for a positive equilibrium of the limiting equations (3.19) for $m = 2$ competing species. Using θ_1 as a bifurcation parameter we see that these equations have a solution if and only if $\theta_1 = \theta_1^{cr}\theta_2/\theta_2^{cr}$, in which case positive solutions are obtained from the equation $u(q_1, q_2) = \theta_1^{cr}/\theta_1$. By the implicit function theorem this equation has a solution $q_1 = q_1(q_2)$, $q_1(q_2^e) = 0$. That is to say, there is a branch of positive equilibria that bifurcates from the equilibrium point $(q_1, q_2) = (0, q_2^e)$ as a function of θ_1 at the critical value $\theta_1 = \theta_1^{cr}\theta_2/\theta_2^{cr}$. But the bifurcation is "vertical" in the sense that the positive equilibria exist only for a single value of θ_1 and it is this "point" spectrum that gives rise to the competitive exclusion principle in this model. (Many competition models in the literature can be viewed in a similar manner.) Thus, in trying to obtain a bifurcation from a stable equilibrium $(0, q_2^e)$ to stable positive equilibria in which both species are present, we find that the spectrum for θ_1 is a single point and hence these positive coexistence equilibria are "nongeneric."

The same bifurcation point of view can be taken when the isolated species q_2 has a stable attractor different from an equilibrium, such as, for example, a stable cycle. If we attempt to find positive coexistence cycles that bifurcate from this stable cycle by using θ_1 as a bifurcation parameter, will the spectrum still be a single point? Or can the bifurcation in this case not be vertical and the spectrum an interval, so that coexistence is possible over a set of θ_1 values that does not have measure 0? The existence of a bifurcating branch of cycles can be proved, and its direction of bifurcation determined, in a manner identical to that used for equilibria by considering composites of the modeling equations (a fixed point of the first composite, that is not an equilibrium, is a 2-cycle; a fixed point of the second composite that is not an equilibrium is a 3-cycle, etc.); see [76]. We will be content here with numerical examples that illustrate such bifurcations.

In Fig. 3.3 graphs of simulations of the limiting equations (3.24) are shown for selected values of the parameters in which species q_2 has a stable 2-cycle in the absence of species q_1. For small values of θ_1 species q_1 goes extinct and species q_2 survives while the opposite is true for large values of θ_1. However, for intermediate values of θ_1 there is a positive stable 2-cycle of the limiting equations which represents an oscillatory coexistence state. The coexistence 2-cycles bifurcate from the 2-cycle on the q_2 axis (which loses its planar stability at the bifurcation point) into the positive quadrant and continue to exist as

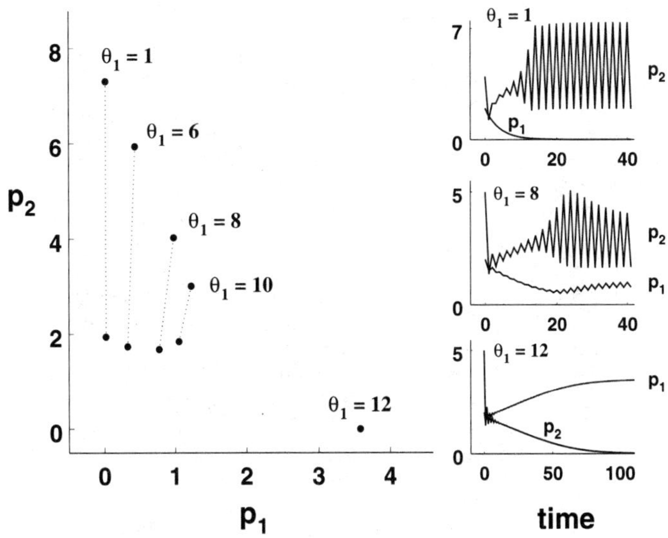

FIG. 3.3. *In the* p_1, p_2 *plane attractors are shown for several values of* θ_1. (*Dotted lines connect* 2-*cycle attractors.*) *The other parameter values in the model equations* (3.24) *are* $\pi_1 = 0.75$, $\pi_2 = 0.25$, $\theta_2 = 100$, $d_1 = d_2 = 1$. *For* $\theta_1 = 1$ *species* p_1 *goes extinct while species* p_2 *tends to a stable* 2-*cycle. For* $\theta_1 = 12$ *species* p_2 *goes extinct while species* p_1 *tends to a stable equilibrium. For intermediate values of* θ_1 *the two species coexist in a stable* 2-*cycle, in contradiction to the fundamental tenet of competitive exclusion* (*which is based upon equilibrium dynamics*).

θ_1 is increased until they are destroyed by a saddle-node bifurcation with the unstable positive equilibrium.

Similar bifurcations into the positive quadrant of higher period cycles can be seen in this model if q_2 has stable cycles of higher periods in the absence of q_1.

Even strange and chaotic attractors can bifurcate into the positive quadrant if q_2 has a strange attractor in the absence of q_1; see Fig. 3.4.

Thus, we see that the classical competitive exclusion principle is very much an equilibrium theory. For situations in which stable nonequilibrium dynamics occur for one of the species it is theoretically possible for two (or more) species to coexist when competing for a single resource, which is in violation of this principle.

An exploitative competition model. Consider the discrete chemostat model in section 1.5.2 in which a micro-organism x is size-structured. This model can be put in the form (3.9) suitable for Theorem 3.1.3 in the following way. Given that $u_{\max} < \nu - 1$ in the Michaelis/Menten uptake function u in (1.64) choose any small number $\varepsilon > 0$ so that $(1+\varepsilon)\, u_{\max} < \nu - 1$ and write the projection matrix (1.63) in the form (3.9) with

$$a(S) = (1 - E)\left(1 - \left(\frac{1+\varepsilon}{\nu - 1}\right) u(S)\right),$$

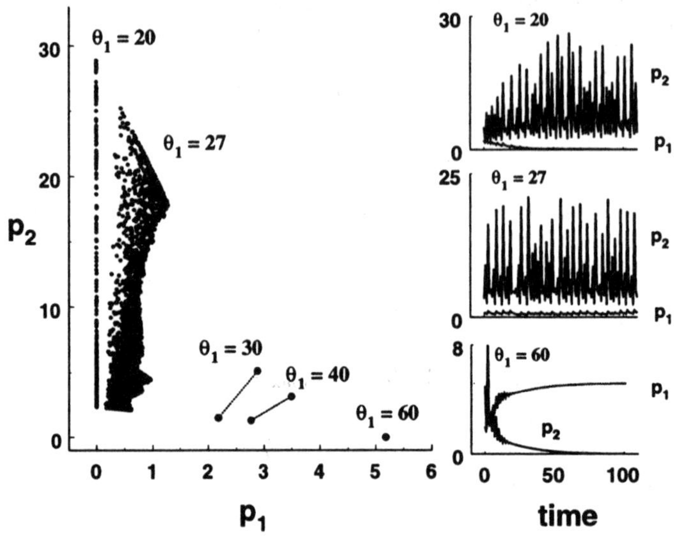

FIG. 3.4. *In the p_1, p_2 plane attractors are shown for several values of θ_1. (Dotted lines connect the two 2-cycle attractors). The other parameter values in the model equations (3.24) are the same as in Fig. 3.3 except that $\theta_2 = 500$. For $\theta_1 = 20$ species p_1 goes extinct while species p_2 tends to a chaotic attractor. For $\theta_1 = 60$ species p_2 goes extinct while species p_1 tends to a stable equilibrium. For intermediate values of θ_1 the two species coexist in a chaotic attractor (at $\theta_1 = 27$) or a stable 2-cycle (at $\theta_1 = 30$ and 40), in contradiction to the fundamental tenet of competitive exclusion.*

$$b(S) = (1 - E)\frac{1}{\nu - 1}u(S),$$

and

$$L = \begin{bmatrix} \varepsilon & 0 & 0 & \cdots & 0 & \nu \\ \nu & \varepsilon & 0 & \cdots & 0 & 0 \\ 0 & \nu & \varepsilon & \cdots & 0 & 0 \\ \vdots & \vdots & \vdots & \ddots & \vdots & \vdots \\ 0 & 0 & 0 & \cdots & \varepsilon & 0 \\ 0 & 0 & 0 & \cdots & \nu & \varepsilon \end{bmatrix}.$$

Here we have relabeled the substrate concentration y in section 1.5.2 as S. The strictly dominant eigenvalue and eigenvectors of L are $r = \nu + \varepsilon$ and

$$v = w = \begin{bmatrix} 1 \\ 1 \\ \vdots \\ 1 \end{bmatrix} > 0.$$

Thus, the population has a stable normalized class distribution and its total biomass $p(t) = \sum_{i=1}^{m_1} x_i(t)$ has the limiting equation (after some simplification)

$$q(t+1) = (1-E)\,(1+u(S))\,q(t). \tag{3.25}$$

This equation, together the limiting equation for the substrate is given by $g(x,S)$ in (1.63) and the Michaelis/Menten expression (1.64) for $u(S)$, yields a planar system of limiting equations for the population level dynamics of the discrete chemostat

$$\begin{aligned} q(t+1) &= (1-E)\left(1+u_{\max}\frac{S(t)}{a+S(t)}\right)q(t), \\ S(t+1) &= (1-E)\left(S(t)-u_{\max}\frac{S(t)}{a+S(t)}q(t)\right)+ES_e. \end{aligned}$$

This system has the extinction equilibrium $(q,S) = (0,S_e)$ for all parameter values. It has a positive equilibrium

$$q = \frac{S_e u_{\max} - E\,(S_e + S_e u_{\max} + a)}{u_{\max} - E(1+u_{\max})}, \qquad S = \frac{aE}{u_{\max} - E(1+u_{\max})} \tag{3.26}$$

if and only if

$$E < E_0 \doteq \frac{S_e u_{\max}}{a + S_e(1+u_{\max})}.$$

The eigenvalues of the Jacobian evaluated at the extinction equilibrium $(q,S) = (0,S_e)$ are the positive numbers

$$(1-E)\left(1+u_{\max}\frac{S_e}{a+S_e}\right), \qquad 1-E,$$

both of which are less than 1 if and only if $E > E_0$.

An equivalent system is obtained if the second equation is replaced by the sum of the equations, namely,

$$\begin{aligned} q(t+1) &= (1-E)\left(1+u_{\max}\frac{S(t)}{a+S(t)}\right)q(t), \\ q(t+1)+S(t+1) &= (1-E)\,[q(t)+S(t)]+ES_e. \end{aligned}$$

The second equation implies that $q(t)+S(t) \to S_e$, and thus, we replace $S(t)$ in the first equation by $S_e - q(t)$ to obtain another limiting equation

$$q(t+1) = (1-E)\left(1+u_{\max}\frac{S_e-q(t)}{a+S_e-q(t)}\right)q(t). \tag{3.27}$$

(This can be viewed as the dynamics on the line segment $q+S = S_e$ lying in the positive quadrant to which all positive solutions tend.) We have reduced the discrete size-structured chemostat model, regardless of its size (i.e., the number of size classes used), to the study of a single scalar map! The limiting equation (3.27) and its mathematical relationship to the dynamics of the original size-structured model are rigorous and thoroughly studied in [386], where it is shown, under appropriate restrictions on the model parameters, that the extinction equilibrium $q = 0$ is globally stable if $E > E_0$ and a unique positive equilibrium is globally stable if $E < E_0$.

If n species are each modeled by a discrete size-structured model of the type above, then each species has a stable normalized size class distribution and each matrix equation has a limiting equation of the form (3.25). The result is a system of $n+1$ limiting equations for the substrate and the total biomasses of the species

$$\begin{aligned} q_i(t+1) &= (1-E)\left(1+u^i_{\max}\frac{S(t)}{a_i+S(t)}\right)q_i(t), \\ S(t+1) &= (1-E)\left(S(t)-\sum_{i=1}^n u^i_{\max}\frac{S(t)}{a_i+S(t)}q_i(t)\right)+ES_e. \end{aligned}$$

Although this system is still rather formidable, it is nonetheless of considerably lower dimension than the original size-structured model. The case of $n = 2$ species is studied in detail in [386]. It is shown there that the familiar chemostat competition result holds, namely, that only one species survives asymptotically and that is the species with the smallest "break-even concentration of nutrient." This break-even concentration is defined to be the level λ_i of substrate at which the coefficient in the equation for q_i is equal to 1, i.e.,

$$\lambda_i = E\frac{a_i}{(1-E)\,u^i_{\max}-E}.$$

3.1.2 Continuous age-structured models. The stable normalized distribution result for the nonlinear discrete model (3.3) has an analog for nonlinear continuous age-structured models [378] (also see [41], [244]). The modeling assumption in the McKendrick age-structured model (2.5) analogous to the scalar multiplicative assumption in the discrete model (3.3) is that the death rate has an additive decomposition into an age specific term plus a density dependent term, namely,

$$\delta(a)+\mu\left(p_1(t),\dots,p_m(t)\right),$$

where p_i is a weighted total population size

$$p_i(t) \doteq \int_0^{a_M}\omega_i(a)\rho(t,a)da$$

as in section 2.2. For this reason the resulting McKendrick age-structured model

$$(3.28)\qquad \begin{aligned} &\text{(a)}\quad \partial_t\rho+\partial_a\rho=-\delta(a)\rho \\ &\qquad\quad -\mu(p_1(t),\dots,p_m(t))\rho,\ t>0,\ 0<a<a_M, \\ &\text{(b)}\quad \rho(t,0)=\textstyle\int_0^{a_M}\beta(a)\rho(t,a)da,\quad t>0, \\ &\text{(c)}\quad \rho(0,a)=\rho_0(a),\quad 0<a<a_M, \\ &\text{(d)}\quad \rho(t,a_M)=0,\quad t>0, \end{aligned}$$

is often called a *separable* model. Biologically the model implies that the effects of density on the death rate act uniformly over all age classes. Note that the fertility rate $\beta(a)$, while age dependent, is density independent in this model. The key to obtaining a stable age distribution result and a limiting equation for total population size

$$p(t) = \int_0^{a_M} \rho(t,a)da$$

is found in an analog to equation (3.5). Straightforward calculations show that the normalized distribution

$$\varphi(t,a) \doteq \frac{\rho(t,a)}{p(t)}$$

of the solutions of *both* the separable model (3.28) and the related linear model (2.9) satisfy the same equations

$$(3.29)\quad \begin{array}{ll} \text{(a)} & \partial_t\varphi + \partial_a\varphi = -\delta(a)\varphi \\ & \quad - \int_0^{a_M} (\beta(\alpha) - \delta(\alpha))\,\varphi(t,\alpha)d\alpha\varphi, \; t>0, \; 0<a<a_M, \\ \text{(b)} & \varphi(t,0) = \int_0^{a_M} \beta(a)\varphi(t,a)da, \quad t>0, \\ \text{(c)} & \varphi(0,a) = \frac{\rho_0(a)}{p(0)}, \quad 0<a<a_M, \\ \text{(d)} & \varphi(t,a_M) = 0, \quad t>0. \end{array}$$

It is here that the additive separability of the density and age-specific terms in the death rate is crucial. It follows that the normalized age distribution of the separable equation has the same stable age distribution as that of the linear equation in section 2.2.1, namely, (2.13) holds.

The nonlinear analog of the linear limiting equation (2.14) (and the continuous analog of the discrete limiting equation (3.11)) can be found by substituting $\rho(t,a) = p(t)\varphi(t,a)$ into equations (3.28) and making use of equations (3.29) and

$$p_i(t) = p(t)\int_0^{a_M} \omega_i(a)\varphi(t,a)da.$$

The result is the ordinary differential equation

$$p' = \left[\int_0^{a_M} (\beta-\delta)\,\varphi d\alpha - \mu\left(p\int_0^{a_M}\omega_1\varphi da, \dots, p\int_0^{a_M}\omega_m\varphi da\right)\right]p$$

for total population size $p = p(t)$. This equation is coupled with the normalized distribution φ. The limiting equation for p is obtained by replacing φ by its asymptotic limit

$$\lim_{t\to+\infty}\varphi(t,a) = \frac{e^{-ra}\Pi(a)}{\int_0^{a_M} e^{-ra}\Pi(a)da},$$

where

$$\Pi(a) \doteq \exp\left(-\int_0^a \delta(\alpha)d\alpha\right)$$

and $r > 0$ is the dominant positive real root of the characteristic equation

$$\int_0^{+\infty} \beta(a)\Pi(a)e^{-rt}da = 1.$$

Using

$$\begin{aligned}\int_0^{a_M} (\beta(\alpha) - \delta(\alpha))\, e^{-r\alpha}\Pi(\alpha)d\alpha &= 1 + \int_0^{a_M} e^{-r\alpha}\frac{d}{d\alpha}\Pi(\alpha)d\alpha \\ &= r\int_0^{a_M} e^{-r\alpha}\Pi(\alpha)d\alpha\end{aligned}$$

we obtain the limiting equation (replacing $p(t)$ by $q(t)$)

$$q' = \left[r - \mu\left(\omega_1^0 q, \dots, \omega_m^0 q\right)\right] q,$$

where

$$\omega_i^0 = \frac{\int_0^{a_M} \omega_i(a)e^{-ra}\Pi(a)da}{\int_0^{a_M} e^{-ra}\Pi(a)da}.$$

Solutions of this scalar autonomous ordinary differential equation have only monotonic dynamics and the only possible asymptotic states for bounded solutions are equilibria.

A special case occurs when the death rate $\mu = \mu(p)$ depends on total population size. In this case, the limiting equation simplifies to the "logisticlike" equation

$$q' = [r - \mu(q)]\, q.$$

Multispecies separable equations result if the death rate of each species is additively separable and the density term depends on the population densities of other species. For example, if for the ith species $\mu_i = \mu_i(p_1, \dots, p_n)$, where p_i is the total population size of the ith species, then the limiting equations for each species form a system of autonomous differential equations

$$q_i' = [r_i - \mu_i(q_1, \dots, q_n)]\, q$$

of the classical (Kolmogorov) type common in theoretical ecology. A linear dependence of μ_i on population sizes result in the famous Volterra system of differential equations.

Rigorous treatments of separable McKendrick age-structured equations can be found in [244] for $a_M < +\infty$ and [41], [378] for $a_M = \infty$. Theorems concerning the existence and uniqueness of solutions, limiting equations, etc. can be found in these references.

3.2 The linear chain trick

If the birth and death rates β and δ in the McKendrick age-structured model

$$\partial_t \rho + \partial_a \rho = -\delta(t,p)\rho, \quad \rho(t,0) = \int_0^\infty \beta(t,p)\rho(t,a)da$$

are independent of age, then an integration of the partial differential equation from $a = 0$ to $+\infty$ (assuming $\lim_{a\to+\infty} \rho(t,a) = 0$) yields the ordinary differential equation

$$p' = [\beta(t,p) - \delta(t,p)]\, p$$

for total population size $p(t)$.

If β and δ depend on age a, but do so in a certain mathematical way, integrations of the partial differential equation can still yield ordinary differential equations for p and/or other weighted total population sizes. The procedure, called the *linear chain trick*, will only be illustrated here by means of selected examples and applications, rather than presented in any general way. Under the right circumstances, it can also be used to derive dynamical equations for population level quantities from models using structuring variables other than age a as well (e.g., see the size-structured competition application below). The linear chain trick relies on a special mathematical type of dependence of the vital birth, death, and/or growth rates on the structuring variable, namely, a polynomial multiplied by an exponential.

The linear chain trick has been extensively applied in the analysis of delay differential equations [86], [303]. Also see [42], [145], [199], [323].

Consider the McKendrick model equations

$$\partial_t \rho + \partial_a \rho = -\delta\rho, \quad \rho(t,0) = \int_0^\infty \beta(a)\rho(t,a)da$$

in which δ is assumed independent of age a and

$$\beta = be^{-ca}. \tag{3.30}$$

For this submodel fertility is an exponentially decreasing function of age. Define the weighted population sizes

$$p_1(t) \doteq \int_0^\infty \rho(t,a)da, \quad p_2(t) \doteq \int_0^\infty e^{-ca}\rho(t,a)da.$$

Then (assuming $\lim_{a\to+\infty} \rho(t,a) = 0$)

$$p_1'(t) = \int_0^\infty \partial_t\rho(t,a)da = -\int_0^\infty [\partial_a\rho(t,a) + \delta\rho(t,a)]\, da$$

and $\rho(t,0) = bp_2(t)$ yield the equation $p_1'(t) = bp_2(t) - \delta p_1(t)$ for $p_1(t)$. An equation for p_2 can be derived by first multiplying the partial differential equation for ρ by e^{-ca} before integrating. Thus,

$$\begin{aligned} p_2'(t) &= \int_0^\infty e^{-ca}\partial_t\rho(t,a)da = -\int_0^\infty e^{-ca}\left[\partial_a\rho(t,a)+\delta\rho(t,a)\right]da \\ &= -\left[e^{-ca}\rho(t,a)\Big|_{a=0}^{a=\infty} + c\int_0^\infty e^{-ca}\rho(t,a)da\right] - \delta p_2(t) \\ &= bp_2(t) - cp_2(t) - \delta p_2(t). \end{aligned}$$

The result is the two-dimensional system

$$\begin{aligned} p_1' &= -\delta p_1 + bp_2, \\ p_2' &= (b-c-\delta)\,p_2 \end{aligned}$$

for the population level quantities p_1 and p_2. If δ and β are constants, this is a linear system for which the origin is (globally asymptotically) stable if and only if the inherent net reproductive number $n = \frac{b}{c+\delta} < 1$. If $n > 1$, both $p_1(t)$ and $p_2(t)$ grow exponentially.

If δ and β are dependent on total population size p_1 (and/or p_2), i.e., if

$$\delta = \delta(p_1, p_2) > 0, \qquad \beta = b(p_1,p_2)e^{-ca} \geq 0, \qquad b(0,0) > 0,$$

then we have the nonlinear plane autonomous system

$$\begin{aligned} p_1' &= -\delta(p_1,p_2)p_1 + b(p_1,p_2)p_2, \\ p_2' &= \left(b(p_1,p_2) - c - \delta(p_1,p_2)\right)p_2. \end{aligned}$$

From a knowledge of p_1 and p_2 the age distribution ρ is known from the formulas[6]

$$\rho(t,a) = \rho_0\,(a-t)\exp\left(-\int_0^t \delta\left(p_1(\sigma),p_2(\sigma)\right)d\sigma\right), \quad 0 \leq t \leq a,$$

$$\rho(t,a) = b\left(p_1(t-a),p_2(t-a)\right)p_2(t-a)\exp\left(-\int_{t-a}^t \delta\left(p_1(\sigma),p_2(\sigma)\right)d\sigma\right), \quad a < t.$$

A biologically more realistic fertility function would vanish for newborns and increase to a maximum at some age $a_m > 0$ before decreasing to 0. For example, we could take

$$\beta = bae^{-ca}, \quad c = \frac{1}{a_m}.$$

[6] A "solution" ρ of the McKendrick model equations gives rise to a solution pair p_1, p_2 of the planar system of ordinary differential equations. Since the converse is not true (the initial conditions for p_1 and p_2 are not arbitrary), the two mathematical problems are not equivalent in the sense that a solution of one yields a solution of the other. The relationship between the asymptotic dynamics of the two systems of equations is studied in [42].

Define

$$\begin{aligned} p_1(t) &\doteq \int_0^\infty \rho(t,a)da, \\ p_2(t) &\doteq \int_0^\infty e^{-ca}\rho(t,a)da, \\ p_3(t) &\doteq \int_0^\infty ae^{-ca}\rho(t,a)da. \end{aligned}$$

Differential equations for these weighted total population sizes can be derived from integrations of the McKendrick partial differential equation (first multiplied by e^{-ca} for p_2 and by ae^{-ca} for p_3). The result is the system

$$\begin{aligned} p_1' &= bp_3 - \delta p_1, \\ p_2' &= bp_3 - (c+\delta)\, p_2, \\ p_3' &= p_2(t) - (c+\delta)\, p_3 \end{aligned}$$

of ordinary differential equations for the p_i.

More generally this procedure can be applied when age-specific fertility is modeled by expressions of the form

$$\beta = ba^k e^{-ca}, \quad c = \frac{k}{a_m}, \quad k \in I[0,+\infty).$$

The integer k inversely measures the "width" of the "reproductive window"; that is to say, for large k fertility is concentrated near the age a_m. Define the weighted population sizes

$$\begin{aligned} p_1(t) &\doteq \int_0^\infty \rho(t,a)da, \\ p_i(t) &\doteq \int_0^\infty a^{i-2}e^{-ca}\rho(t,a)da, \qquad i \in I[2,k+2]. \end{aligned} \tag{3.31}$$

Multiplying the McKendrick partial differential equation by $a^i e^{-ca}$ and integrating the result from $a = 0$ to $+\infty$, we obtain a system of ordinary differential equations for the weighted population sizes $p_i(t)$

$$\begin{aligned} p_1' &= bp_{k+2} - \delta p_1, \\ p_2' &= bp_{k+2} - (c+\delta)\, p_2, \\ p_i' &= (i-2)p_{i-1} - (c+\delta)\, p_i, \quad i \in I[3,k+2]. \end{aligned} \tag{3.32}$$

The derivation of this system of ordinary differential equations remains valid when $\delta = \delta(p_i)$ and $b = b(p_i)$, in which case the system becomes a nonlinear system of ordinary differential equations for the $k+2$ weighted population sizes p_i. The age distribution is given by

$$\rho(t,a) = \rho_0(a-t)\exp\left(-\int_0^t \delta\left(p_i(\sigma)\right)d\sigma\right), \quad 0 \le t \le a,$$

$$\rho(t,a) = b\left(p_i(t-a)\right) p_{k+2}(t-a) \exp\left(-\int_{t-a}^{t} \delta\left(p_i(\sigma)\right) d\sigma\right), \quad a < t.$$

We consider two applications in which the linear chain trick is used. The first involves an age-structured single species model. The second involves a size-structured multispecies competition model. For an application of the linear chain trick to age-structured predator–prey models see [201].

Age-specific fertility windows. In order to investigate some effects of age dependent fertility we consider the McKendrick model with submodels

$$\begin{array}{l} \delta = \text{constant} > 0, \quad \beta = b(p_1)a^k e^{-ca}, \quad c = \frac{k}{a_m}, \quad k \in I[0,+\infty), \\ b(0) > 0, \quad b'(p) < 1, \quad \lim_{p\to+\infty} b(p) = 0. \end{array}$$

Here $p_1(t) \doteq \int_0^{+\infty} \rho(t,a)da$ is total population size. The focus is on age dependent fertility, so age dependence in mortality is ignored. The birth rate is dependent both on age and on total population size p_1. At any population size p_1 maximum fertility occurs at age a_m and as k gets larger fertility becomes more concentrated near this age.

The Jacobian of the differential system (3.32) for the weighted population sizes p_i (3.31) is, at the extinction equilibrium $p_i = 0$,

$$\begin{bmatrix} -\delta & 0 & 0 & \cdots & 0 & b(0) \\ 0 & -(c+\delta) & 0 & \cdots & 0 & b(0) \\ 0 & 1 & -(c+\delta) & \cdots & 0 & 0 \\ 0 & 0 & 2 & \cdots & 0 & 0 \\ \vdots & \vdots & \vdots & \ddots & \vdots & \vdots \\ 0 & 0 & 0 & \cdots & -(c+\delta) & 0 \\ 0 & 0 & 0 & \cdots & k & -(c+\delta) \end{bmatrix}.$$

The eigenvalues of this Jacobian are

$$z = -\delta, \qquad z = \left(k!b(0)\right)^{1/(k+1)} - (c+\delta).$$

Thus, the extinction equilibrium is (locally asymptotically) stable if the inherent net reproductive number

$$n = \int_0^{+\infty} \beta(a,0)e^{-\delta a}da = k!b(0)\,(c+\delta)^{-k-1}$$

satisfies $n < 1$ and unstable if $n > 1$.

There is a positive equilibrium if (and only if)

$$n = k!b(0)\,(c+\delta)^{-k-1} > 1,$$

namely,

$$\begin{aligned} p_1 &= b^{-1}\left(\frac{(c+\delta)^{k+1}}{k!}\right), \\ p_{i-1} &= \frac{c+\delta}{i-2}p_i, \quad i=3,\dots,k+2, \\ p_{k+2} &= \delta\frac{k!}{(c+\delta)^{k+1}}p_1. \end{aligned}$$

We know from the general bifurcation results in section 2.2 that this positive equilibrium is (locally asymptotically) stable for n close to 1. In fact this equilibrium is stable for all $n > 1$. To see this, we calculate the Jacobian of the differential system (3.32) at the positive equilibrium and obtain

$$\begin{bmatrix} b'(p_1)p_{k+2}-\delta & 0 & 0 & \cdots & 0 & b(p_1) \\ b'(p_1)p_{k+2} & -(c+\delta) & 0 & \cdots & 0 & b(p_1) \\ 0 & 1 & -(c+\delta) & \cdots & 0 & 0 \\ 0 & 0 & 2 & \cdots & 0 & 0 \\ \vdots & \vdots & \vdots & \ddots & \vdots & \vdots \\ 0 & 0 & 0 & \cdots & k & -(c+\delta) \end{bmatrix}.$$

The characteristic equation

$$(\zeta - b'(p_1)p_{k+2}+\delta)\left((\zeta+c+\delta)^{k+1} - k!b(p_1)\right) - k!b(p_1)b'(p_1)p_{k+2} = 0$$

can be rewritten as

$$\frac{\zeta - b'(p_1)p_{k+2}+\delta}{\zeta+\delta} = \left(\frac{c+\delta}{\zeta+c+\delta}\right)^{k+1}.$$

For any complex number with $\operatorname{Re}\zeta \geq 0$ the magnitude of the right-hand side is less than 1 while that of the left-hand side is strictly greater than 1. Thus, there is no root (eigenvalue) with $\operatorname{Re}\zeta \geq 0$ and the equilibrium is (locally asymptotically) stable.

An interesting conclusion suggested by this example is that age-dependent fertility is not destabilizing no matter how narrowly defined the reproductive window is or how late the "maturation period delay" a_m is. See [85] for a more general consideration of this assertion.

Size-structured competition. In [94] and [96] a competition model for size-structured species is considered (also see [387]). One motivation for considering size-structured species comes from the question of whether body size confers any competitive advantage to a species. See [129], [148], and [212] (and the references therein) concerning this issue.

The model in [94] is for exploitative competition among species; that is to say, individuals do not directly interfere with each other in their competition for the limiting resource, but only compete indirectly through their mutual consumption of the resource. Mathematically, this means the equations for the species dynamics are not coupled to one another. For example, competition in a chemostat is usually considered exploitative [387].

The abundance (concentration) of the limiting resource is $R = R(t)$ and its dynamics are assumed governed by an ordinary differential equation $R' = f(R)$ in the absence of the species. Prototypical resource dynamics are the "chemostat" model $f(R) = r(R_0 - R)$ and the "renewable resource (logistic)" model $f(R) = r\left(1 - \frac{R}{K}\right)R$.

Each size-structured species is modeled by equations of the form (2.30), i.e.,

$$\begin{aligned} \partial_t \rho + \partial_s(\gamma\rho) &= -\delta\rho, \quad t > 0, \quad s_b < s < +\infty, \\ \gamma\rho|_{s=s_b} &= \int_{s_b}^{+\infty} b\rho(t,s)ds, \quad t > 0, \end{aligned}$$

where s is some measure of body size (to be specified below). All newborns are assumed to have the same size s_b at birth. The submodels for the vital rates b and γ are based on the assumption that growth, reproductive, and metabolic rates scale exponentially to body length l [435]. We will assume, for simplicity, that the death rate δ is not size specific, i.e., $\delta =$ constant > 0.

Birth and growth rates depend on the resource uptake rate. It is assumed that the resource uptake rate $u(R)l^\mu$ scales to body length l and that some portion of the consumed resource is utilized for metabolism, leaving a net amount $n(R)l^\mu$ available for growth and reproduction [212]. If w denotes an individual's weight (assumed proportional to volume l^3), then weight change is proportional to $n(R)l^\mu$, i.e.,

$$\frac{dw}{dt} = \frac{\kappa}{\eta} n(R) l^\mu,$$

where $\kappa \in (0,1)$ is the fraction of consumed resource that is allocated to growth (and $1-\kappa$ is the fraction allocated to reproduction) and η is a conversion factor relating weight to resource units. Then since $\gamma = \frac{ds}{dt}$, we have

$$\gamma = \frac{\kappa}{\eta} n(R) l^\mu \frac{ds}{dw}$$

and

$$b = \frac{1-\kappa}{\xi w_b} n(R) l^\mu,$$

where ξ is a resource-to-offspring conversion factor and w_b is the weight of newborns. (Notice the simplifying assumption has been made that individuals of all sizes $s > s_b$ are reproductive.) The coefficients

$$\alpha = \frac{1-\kappa}{\xi}, \qquad \beta = \frac{\kappa}{\eta}$$

are the *reproductive and growth efficiency coefficients*, respectively.

We are interested in the competition of m species for the resource R. Therefore the quantities in the model equations above are subscripted by $i \in I[1, m]$. Let s_i denote the size s_b at birth for the ith species. The model equations for m size-structured species are

$$
\begin{aligned}
\partial_t \rho_i + \partial_s (\gamma_i \rho_i) &= -\delta_i \rho, \quad t > 0, \quad s_i < s < +\infty, \\
\gamma_i \rho_i |_{s=s_i} &= \int_{s_i}^{+\infty} b_i \rho_i(t, s) ds, \quad t > 0, \\
R' &= f(R) - \sum_{i=1}^{m} u_i(R) \int_{s_i}^{+\infty} s \rho_i(t, s) ds.
\end{aligned}
$$

In [96] the cases $\mu = 3$ and $\mu = 2$ are considered where resource consumption scales to body volume and to body surface area, respectively. In the first case the structuring variable is taken to be body volume so that $s = l^3$. In the second case $s = l$. Both cases are amenable to the linear chain trick analysis. We will consider here only what turns out to be the mathematically simpler case $\mu = 3$. See [94] or [387] for the details of the case $\mu = 2$.

For $\mu = 3$ the model equations become

$$
\begin{aligned}
\partial_t \rho_i + \beta_i n_i(R) \partial_s (s \rho_i) &= -\delta_i \rho_i, \quad t > 0, \quad s_i < s < +\infty, \\
\beta_i n_i(R) s \rho_i |_{s=s_i} &= \frac{\alpha_i}{s_i} n_i(R) \int_{s_i}^{+\infty} s \rho_i(t, s) ds, \quad t > 0, \\
R' &= f(R) - \sum_{i=1}^{m} u_i(R) \int_{s_i}^{+\infty} s \rho_i(t, s) ds.
\end{aligned}
$$

Differential equations for the total population sizes and the total population volumes

$$
p_i(t) = \int_{s_i}^{+\infty} \rho_i(t, s) ds, \qquad v_i(t) = \int_{s_i}^{+\infty} s \rho_i(t, s) ds
$$

can be derived as in the age-structured examples above, namely, by integrating the partial differential equation from $s = s_i$ to $+\infty$ (multiplying the equation by s in the case of v_i). The result is the system of ordinary differential equations

$$
\begin{aligned}
p_i' &= -\delta_i p_i + \frac{\alpha_i}{s_i} n_i(R) v_i, \\
v_i' &= (-\delta_i + (\alpha_i + \beta_i) n_i(R)) v_i, \\
R' &= f(R) - \sum_{i=1}^{m} u_i(R) v_i.
\end{aligned}
$$

Since the first of these equations is uncoupled from the remaining equations we can determine the asymptotic dynamics from the system

$$
\begin{aligned}
v_i' &= (-\delta_i + (\alpha_i + \beta_i) n_i(R)) v_i, \\
R' &= f(R) - \textstyle\sum_{i=1}^{m} u_i(R) v_i.
\end{aligned}
\tag{3.33}
$$

Total population numbers are then determined from the variation of constants formula

$$p_i(t) = \exp(-\delta_i t)\left(p_i(0) + \frac{\alpha_i}{s_i}\int_0^t n_i(R(s))v_i(s)\exp(\delta_i s)\,ds\right).$$

A straightforward calculation of the derivative of the average volume

$$\bar{v}_i \doteq \frac{v_i}{p_i}$$

shows that

$$\bar{v}_i' = n_i(R)\left(\alpha_i + \beta_i - \frac{\alpha_i}{s_i}\bar{v}_i\right)\bar{v}_i.$$

Thus, on the time scale $\tau_i \doteq \int_0^t [n_i(R(v))]^{-1}\,dv$, the average volume of an individual $\bar{v}_i$ satisfies the famous logistic equation

$$\bar{v}_i' = \left(\alpha_i + \beta_i - \frac{\alpha_i}{s_i}\bar{v}_i\right)\bar{v}_i$$

and hence equilibrates to

$$\lim_{t\to+\infty} \bar{v}_i = \bar{v}_i^\infty \doteq \frac{\alpha_i + \beta_i}{\alpha_i}s_i > s_i.$$

This limiting average volume $\bar{v}_i^\infty$ for individuals of the species is independent of the asymptotic dynamics of the species as a component of the system (3.33) and, in particular, is independent of whether the species goes extinct or survives, equilibrates or oscillates! We define $\bar{v}_i^\infty$ to be the "size" of the species. This "species size" is related to the original model parameters by the formula

$$\bar{v}_i^\infty = \frac{\eta_i(1-\kappa_i) + v_i\kappa_i}{\eta_i(1-\kappa_i)}s_i.$$

The outcome of the competition described by the system (3.33) depends on the designated submodels for the resource dynamics $f(R)$ and for the uptake rates $u_i(R)$ and $n_i(R)$. Most submodels will satisfy the minimal conditions

$$\text{(3.34)}\qquad \begin{array}{l} u_i,\ n_i \in C^1\left(R_+^1, R_+^1\right), \quad u_i(0) = n_i(0) = 0, \quad u_i(R) \geq n_i(R), \\ f \in C^1\left(R_+^1, R_+^1\right), \quad f(0) \geq 0. \end{array}$$

First of all, we note that the positive cone R_+^{m+1} is forward invariant under the flow defined by (3.33). That $v_i(0) > 0$ implies that $v_i(t) > 0$ for all t follows from

$$v_i(t) = v_i(0)\exp\left(-\delta_i t + (\alpha_i + \beta_i)\int_0^t n_i(R)ds\right).$$

Suppose that $R(0) > 0$. If $f(0) = 0$, then

$$R(t) = R(0) \exp \left[\int_0^t \left(\frac{f(R)}{R} - \sum_{i=1}^{m} \frac{u_i(R)}{R} v_i \right) ds \right] > 0$$

for all t. On the other hand, if $f(0) > 0$ and if $t_1 > 0$ were the first time at which $R(t)$ vanishes, then $0 \geq R'(t_1) = f(0)$, a contradiction that implies that no such first time t_1 exists.

Second, if we assume that

(3.35) solutions of $R' = f(R)$, $R(0) > 0$, are bounded for $t > 0$,

then we can argue that positive solutions of (3.33) remain bounded for $t > 0$. That $R(t)$ is bounded for $t > 0$ follows from $R' \leq f(R)$ and standard comparison theorems. To show that $v_i(t)$ is bounded, define $v = R + \sum_{i=1}^{m} (\alpha_i + \beta_i)^{-1} v_i$ and obtain, setting $\delta = \min \delta_i > 0$,

$$\begin{aligned} v' &= f(R) + \sum_{i=1}^{m} (n_i(R) - u_i(R))\, v_i - \sum_{i=1}^{m} (\alpha_i + \beta_i)^{-1} \delta_i v_i \\ &\leq f(R) - \delta \sum_{i=1}^{m} (\alpha_i + \beta_i)^{-1} v_i \end{aligned}$$

or finally the inequality

$$v' \leq -\delta v + f(R) + \delta R.$$

In as much as R, and hence $f(R)$, have been shown to be bounded for $t > 0$ it follows that $v(t)$ is bounded for $t > 0$. Since v_i is known to be positive, it follows from the definition of v that each v_i is bounded for $t > 0$.

In summary, under the assumptions (3.34) and (3.35) all solutions of (3.33) with $R(0) > 0$ and $x_i(0) > 0$ are positive and bounded for $t > 0$.

Consider the case when the resource dynamics are given by

$$f(R) = r(R_0 - R), \quad r > 0.$$

With this submodel for f, (3.33) is a model for micro-organisms growing in a chemostat [387]. In this case, the resource equilibrates exponentially to R_0 in the absence of the micro-organism populations from the chemostat. Let us assume a monotonic uptake rate, $u_i'(R) > 0$. The so-called Michaelis/Menten (or Holling II or Monod) uptake rate $u = u_0 \frac{R}{a+R}$ is the most famous example. Furthermore, we assume that the net uptake rate n_i is a fixed fraction ψ_i of the total uptake rate, so that $n_i(R) = \psi_i u_i(R)$. Finally, we assume that $\delta_i = r$ for all i. This latter assumption, almost always made in chemostat models, means that the removal of micro-organisms per unit time due to the washout rate r in the chemostat far exceeds the mortality rate of the micro-organisms per unit time (which is then ignored in the model). The system (3.33) now becomes

$$\begin{aligned} v_i' &= (-r + (\alpha_i + \beta_i)\, \psi_i u_i(R))\, v_i, \\ R' &= r(R_0 - R) - \textstyle\sum_{i=1}^{m} u_i(R) v_i. \end{aligned} \tag{3.36}$$

This competition system has been well studied [387], [43]. It is known to have global equilibrium dynamics only and that at most one species can survive (in keeping with the competitive exclusion principle). The asymptotic dynamics are determined by the so-called *break-even concentrations* λ_i defined as the (unique) solution of the equation

$$-r + (\alpha_i + \beta_i)\,\psi_i u_i(\lambda) = 0.$$

All species will die out except the one with the smallest break-even value λ_i (provided it exceeds a minimum threshold value, namely, the input concentration R_0 of the limiting resource). Thus, if $\lambda_k = \min \lambda_i$, then

$$\lambda_k > R_0 \text{ implies that } \begin{cases} \lim_{t\to+\infty} R(t) = R_0, \\ \lim_{t\to+\infty} v_i = \lim_{t\to+\infty} p_i = 0 \text{ for all } i \end{cases}$$

and

$$\lambda_k < R_0 \text{ implies that } \begin{cases} \lim_{t\to+\infty} R(t) = R_\infty < R_0, \\ \lim_{t\to+\infty} v_i = \lim_{t\to+\infty} p_i = 0 \text{ for all } i \neq k, \\ \lim_{t\to+\infty} v_k = v_\infty > 0,\ \ \lim_{t\to+\infty} p_k = p_\infty > 0. \end{cases}$$

In the second case

$$R_\infty = \lambda_1, \quad v_\infty = (R_0 - \lambda_k)\,(\alpha_k + \beta_k)\,, \qquad p_\infty = \frac{\alpha_k}{s_k}\,(R_0 - \lambda_k)\,,$$

$$\alpha_k = \frac{1-\kappa_k}{\xi_k}, \qquad \beta_k = \frac{\kappa_k}{\eta_k}$$

defines the equilibrium state of the surviving species.

How does species size $\bar{v}_i^\infty$ relate to competitive efficiency? Does the winning species, i.e., the species with the smallest break-even concentration λ_i, have the largest species size $\bar{v}_i^\infty$? The answer is that in this model there is no simple relationship between species size and competitive success, at least not without further restrictive assumptions about the species involved. For example, consider m similar species, similar in the sense that they all have the same resource uptake rate $u_i(R) = u(R)$ and the same metabolic demands $\psi_i = \psi$. In this case the species with largest value of $\alpha_i + \beta_i$ will have the smallest break-even concentration

$$\lambda_i = u^{-1}\left(\frac{r}{(\alpha_i + \beta_i)\,\psi}\right).$$

However, the species with the largest $\alpha_i + \beta_i$ is not necessarily the species with the largest species size

$$\bar{v}_i^\infty = \frac{\alpha_i + \beta_i}{\alpha_i} s_i.$$

For more discussion of this issue see [94]. Similar conclusions are obtained in the case $\mu = 2$ for which the model equations are

$$\begin{aligned}
\partial_t \rho_i + \frac{\beta_i}{3} n_i(R) \partial_s \rho_i &= -\delta_i \rho_i, \quad t > 0, \quad s_i < s < +\infty, \\
\frac{\beta_i}{3} n_i(R) \rho_i \Big|_{s=s_i} &= \frac{\alpha_i}{s_i^3} n_i(R) \int_{s_i}^{+\infty} s^2 \rho_i(t,s) ds, \quad t > 0, \\
\frac{dR}{dt} &= f(R) - \sum_{i=1}^{m} u_i(R) \int_{s_i}^{+\infty} s^2 \rho_i(t,s) ds.
\end{aligned}$$

3.3 Hierarchical models

For those models considered so far in this chapter the effects of population density have been modeled by assuming that vital rates are functions of one or more weighted total population sizes of the forms

$$\sum_{i=1}^{m} \omega_i x_i(t) \quad \text{or} \quad \int_0^{a_M} \omega(\alpha) \rho(t,\alpha) d\alpha$$

for discrete and continuous models, respectively. A more general situation is one in which the effects of population density that are felt by an individual are dependent on that individual's class (age, size, etc.), so that the weights ω_i (or $\omega(\alpha)$) depend on the structuring variable. Weights ω_{ij} (or $\omega(\alpha, a)$) then measure the effect that individuals of class i (or α) have on the vital rates of individuals of class j (or a). One situation in which this is the case occurs when there is a hierarchy established by the structuring variable that determines an individual's access to resources.

For example, *contest competition* can be defined as the situation when no individual in a class of lower rank (say, of lower age or smaller size) can affect the amount of resource available to an individual of greater rank (of greater age or size). *Scramble competition*, on the other hand, can be defined as the opposite extreme when every individual can affect the amount of resource available to any other individual in the population [297]. In these circumstances the vital rates in a structured model would depend on functionals of the form

$$\sum_{i=1}^{j} \omega_i x_i(t) \quad \text{or} \quad \sum_{i=j}^{m} \omega_i x_i(t)$$

in discrete models and

$$\int_0^{a} \omega(\alpha) \rho(t,\alpha) d\alpha \quad \text{or} \quad \int_a^{a_M} \omega(\alpha) \rho(t,\alpha) d\alpha$$

in continuous models. We call such models *hierarchical models*.

For both discrete and continuous hierarchical models, it turns out to be possible to derive dynamical equations for total population size.

3.3.1 Continuous age-structured models. In a McKendrick age-structured model with $\alpha_M = +\infty$ the integrals

$$Y(t,a) \doteq \int_0^a \rho(t,\alpha)d\alpha, \qquad O(t,a) \doteq \int_a^\infty \rho(t,\alpha)d\alpha$$

give the number of individuals of age less than a and the number of individuals of age greater than a, respectively. In a hierarchical McKendrick model the fertility and death rates are functions of Y and O, so that

$$\begin{aligned} \partial_t\rho + \partial_a\rho &= -\delta(t,Y(t,a),O(t,a))\rho, \quad t>0, \quad 0<a<+\infty, \\ \rho(t,0) &= \int_0^\infty \beta(t,Y(t,a),O(t,a))\rho(t,a)da, \quad t>0, \\ \rho(0,a) &= \rho_0(a), \quad 0<a<+\infty. \end{aligned} \tag{3.37}$$

Note that the total population size is given by

$$p(t) = Y(t,a) + O(t,a).$$

These models are not of the form (2.6) studied in section 2.2. However, models of the form (3.37) are amenable to considerable analysis by means of a single ordinary differential equation for $p(t)$. Heuristically, this equation can be derived as follows. Define

$$B(t,p) \doteq \int_0^p \beta(t,z,p-z)dz, \quad D(t,p) \doteq \int_0^p \delta(t,z,p-z)dz.$$

Noting that $\partial_a Y(t,a) = \rho(t,a)$ we obtain

$$\begin{aligned} \int_0^\infty \beta(t,Y(t,a),O(t,a))\rho(t,a)da &= B(t,p(t)), \\ \int_0^\infty \delta(t,Y(t,a),O(t,a))\rho(t,a)da &= D(t,p(t)). \end{aligned}$$

These equalities, together with an integration of the differential equation in (3.37) from $a=0$ to $+\infty$, lead to the scalar ordinary differential equation

$$p' = B(t,p) - D(t,p) \tag{3.38}$$

for $p = p(t)$. The relationship between this equation and the model equations (3.37) is rigorously studied in [104], where it is shown, under certain conditions, that a solution $\rho(t,a)$ of (an appropriately formulated version of) the initial value problem for (3.37) defines a solution $p(t) = \int_0^{+\infty}\rho(t,\alpha)d\alpha$ of (3.38) and vice versa. These conditions are

$$\begin{aligned} &\beta(\cdot,\cdot,\cdot),\ \delta(\cdot,\cdot,\cdot) \in C^0\left(R^3_+, R^1_+\right), \\ &\beta(t,\cdot,\cdot),\ \delta(t,\cdot,\cdot) \in C^1\left(R^2_+, R^1_+\right),\ t \in R^1_+, \\ &\delta(t,x,y) \ge \delta_0 > 0,\ 0 \le \beta(t,x,y) \le \beta_0 < +\infty,\ (t,x,y) \in R^3_+. \end{aligned} \tag{3.39}$$

It is also shown in [104] that the initial value problem for (3.37) is well posed under these conditions.

When $\delta = \delta(Y, O)$ and $\beta = \beta(O, Y)$ are independent of time t, then (3.38) is an autonomous, scalar ordinary differential equation and therefore has only monotonic dynamics. Thus, the asymptotic dynamics at the population level are easily analyzed by means of the real valued function $B(p) - D(p)$ of the real variable p.

If $p(t)$ approaches an equilibrium p_∞ as $t \to +\infty$, the dynamics of $Y(t, a)$ and $\rho(t, a)$ can be deduced as follows. From the definition of Y and (3.37) follows

$$\partial_t Y + \partial_a Y = B(p) - \int_0^Y \delta\,(z, p - z)\,dz, \quad Y(t, 0) = 0,$$

or in coordinates $(\tau, \alpha) = (t - a, a)$

$$\frac{dY}{d\alpha} = B(p(\tau + \alpha)) - \int_0^Y \delta(z, p(\tau + \alpha) - z)dz.$$

This is a scalar ordinary differential equation satisfied by Y as a function of α with τ as a parameter. Classical theorems for the continuous dependence of solutions on parameters imply that $\lim_{t\to+\infty} Y(t, a) = y_\infty(a)$ uniformly for $\alpha = a$ restricted to a compact interval where $y_\infty(a)$ is the solution of the scalar ordinary differential equation

$$y' = B(p_\infty) - \int_0^y \delta(z, p_\infty - z)dz, \quad y(0) = 0,$$

or equivalently

$$y' = \int_y^{p_\infty} \delta(z, p_\infty - z)dz, \quad y(0) = 0.$$

Note that $\delta > 0$ implies that $\lim_{a\to+\infty} y_\infty(a) = p_\infty$. With regard to $\rho(t, a)$, for $t > a$

$$\rho(t, a) = B(p(t)) \exp\left(-\int_0^a \widehat{\delta}(t - a + \alpha, \alpha)d\alpha\right),$$

where

$$\widehat{\delta}(t, a) \doteq \delta(Y(t, a), p(t) - Y(t, a)).$$

From $\lim_{t\to+\infty} y(t) = p_\infty$ and $\lim_{t\to+\infty} Y(t, a) = y_\infty(a)$ it follows that $\rho(t, a)$ approaches a limit as $t \to +\infty$, uniformly for a restricted to any compact interval, and this limit is given by the formula

$$\rho_\infty(a) \doteq B(p_\infty) \exp\left(-\int_0^a \delta\,(y_\infty(a), p_\infty - y_\infty(a))\,da\right). \tag{3.40}$$

THEOREM 3.3.1. *Consider the autonomous age-structured McKendrick model* (3.37) *where* $\delta = \delta(Y, O)$ *and* $\beta = \beta(Y, O)$ *satisfy* (3.39). *Then the total population size* $p(t)$ *satisfies the scalar ordinary differential equation*

$$p' = B(p) - D(p), \quad p(0) = \int_0^\infty \rho_0(a)da,$$

$$B(p) \doteq \int_0^p \beta(z, p-z)dz, \quad D(p) \doteq \int_0^p \delta(z, p-z)dz$$

and consequently is monotonic in time t. *If* $p(t)$ *asymptotically approaches an equilibrium* p_∞, *then* $\lim_{t\to+\infty} \rho(t,a) = \rho_\infty(a)$ *uniformly for* a *restricted to any compact interval where* $\rho_\infty(a)$ *is defined by* (3.40).

Contest versus scramble competition. One problem of interest concerns the relative advantages or disadvantages of different types of intraspecific competition among the members of a population. Lomnicki [297] argues that contest competition is generally more advantageous and is therefore expected to be more common in nature. If the competition hierarchy is based upon chronological age (or at least correlates closely with chronological age), then one approach to the study of contest competition is to utilize the hierarchical McKendrick model (3.37) and the results in Theorem 3.3.1. Two issues are involved in addressing Lomnicki's tenet about the advantage of contest over scramble competition: how are these two fundamental types of competition modeled using (3.37) and how is an appropriate comparison made between the two models?

Let R denote the amount of a limiting resource available for consumption by an age-structured population. Let c be the fraction of this amount that is available to an individual. In the presence of competition this fraction c will be dependent on population density in some way. A reasonable assumption for age-structured contest competition, in which older individuals are dominant, is that for an individual of age a the fraction c is a decreasing function of the number of older individuals, i.e., $c = c_c(O)$, where $O = O(t,a)$. This assumes that younger individuals have no effect on an individual's resource availability. (If the hierarchy is such that younger individuals are dominant, then c is a decreasing function of $Y = Y(t,a)$.) For scramble competition the fraction c could be taken as a decreasing function $c = c_s(p)$ of total population size $p = p(t)$. Thus, under contest competition the amount of resource available to an individual of age a is $Rc_c(O)$, while under scramble competition the amount is $Rc_s(p)$. If u is the resource uptake rate, then for scramble competition we have a resource uptake rate $u_s = u(Rc_s(p))$ and for contest competition we have a resource uptake rate $u_c = u(Rc_c(O))$.

The next modeling assumptions provide submodels for the vital rates δ and β in the McKendrick model (3.37). We concentrate in this example on the effects that resource competition have on fertility. Therefore, we make the simplifying assumption that the death rate is a constant $\delta = \delta_0 > 0$. With regard to fertility, we assume that the birth rate β is proportional to the resource uptake rate. Thus,

$$\text{scramble competition:} \quad \beta = \beta_0 u(Rc_s(p)),\ \beta_0 > 0,\ \delta = \delta_0 > 0,$$

$$\text{contest competition:} \quad \beta = \beta_0 u(Rc_c(O)),\ \beta_0 > 0,\ \delta = \delta_0 > 0,$$

where

$$u \in C^2\left(R_+^1, R_+^1\right), \qquad u' > 0, \qquad u(0) = 0, \tag{3.41}$$

and both competition fractions c_s and c_c satisfy

$$c \in C^1\left(R_+^1, [0,1]\right), \qquad c' < 0, \qquad c(0) = 1, \qquad \lim_{z \to +\infty} c(z) = 0. \tag{3.42}$$

In order that the total amount of resource available to the whole population remains less that R, i.e., that $\int_0^\infty Rc\rho da \leq R$, the competition fractions should satisfy

$$c_s(p)p \leq 1, \qquad \int_0^p c_c(z)dz \leq 1 \quad \text{for all } p > 0.$$

With these designations of the submodels in (3.37) we have the models for intraspecific scramble and contest competition considered in [230].

In order to make an appropriate comparison between the two types of competition, we impose the criterion that for both types the same amount of resource is utilized (for a given density function $\rho(t,a)$). Thus, we require

$$\int_0^\infty Rc_s(p(t))\rho(t,a)da = \int_0^\infty Rc_c(O(t,a))\rho(t,a)da.$$

If R is independent of age a, this requires the relationship

$$c_s(p) = \frac{1}{p}\int_0^p c_c(z)dz$$

between the competition fractions c_s and c_c.

Given that both types of competition are utilizing the same total amount of resource, which mode is more "advantageous"? What is meant by "advantageous"? We consider two criteria for comparing the two types of competition: equilibrium level and equilibrium resilience. Assuming the equilibrium to be stable, resilience is defined to be the smallest magnitude of the real parts of all eigenvalues of the Jacobian; the resilience provides a lower bound for the rate of approach to the equilibrium. Larger resilience values mean a "more stable" equilibrium in that the return to equilibrium from small perturbations is faster.

By Theorem 3.3.1 the dynamics of total population size, for scramble and for contest competition, are governed by the ordinary differential equations

$$\begin{aligned} &\text{scramble competition:} \quad p' = \delta_0\left(nf_s(p) - 1\right)p, \\ &\text{contest competition:} \quad p' = \delta_0\left(nf_c(p) - 1\right)p, \end{aligned} \tag{3.43}$$

respectively, where

$$f_s(p) \doteq \frac{1}{u(R)} u\left(\frac{1}{p}\int_0^p Rc(z)dz\right),$$
$$(3.44)$$
$$f_c(p) \doteq \frac{1}{u(R)}\frac{1}{p} u\left(\int_0^p Rc(z)dz\right),$$

and where c satisfies (3.42). The quantity

$$n \doteq \frac{\beta_0}{\delta_0} u(R)$$

is the inherent net reproductive number. From the assumptions placed on u and c it follows that both f_s and f_c equal 1 at $p = 0$ and decrease monotonically to 0 as $p \to +\infty$. It is straightforward to deduce from these facts that each of the scalar, autonomous ordinary differential equations in (3.43) has a unique positive equilibrium for $n > 1$ and that this equilibrium is globally asymptotically stable (for initial conditions $p(0) > 0$). For $n < 1$ the extinction equilibrium $p = 0$ is globally asymptotically stable (for initial conditions $p(0) > 0$) for both equations.

Consider the case $n > 1$, and let $p_s > 0$ and $p_c > 0$ denote the globally stable positive equilibria of the scramble and contest competition models (3.43), respectively. If $u''(z) < 0$ for $0 < z < R$, it follows from Jensen's Inequality that $f_c(p) < f_s(p)$ for all $p > 0$ (see [230]). Since $f_c(p_c) = n = f_s(p_s)$ and since f_c and f_s are decreasing functions, it follows that $p_c < p_s$. A similar argument shows that if $u''(z) > 0$ for $0 < z < R$, then $p_c > p_s$. This result shows that the concavity properties of the nonlinearity in the resource uptake rate determines which of the two types of competition has the higher equilibrium level.

For example, the Holling II (Michaelis/Menten or Monod) uptake rate

$$u(z) = u_0 \frac{z}{a+z}$$

satisfies $u''(z) < 0$ for all $z > 0$. Thus, for this model scramble competition is more advantageous than contest competition in the sense that a higher total population equilibrium size is attained under scramble competition. In the case of a Holling III uptake rate

$$u(z) = u_0 \frac{z^2}{a+z^2}$$

there is a change in concavity in u as a function of z. For low resource levels R, $u''(z) > 0$ for $0 < z < R$ and contest competition leads to a higher equilibrium level. For larger resource levels it is possible that this can be reversed (see [230] for an example).

A comparison can also be made of the equilibrium age densities under scramble and contest competition. These densities are $\delta_0 p_s e^{-\delta_0 a}$ and $\delta_0 p_c e^{-\delta_0 a}$, respectively, which show that whatever relationship is borne between the population level equilibria is also borne by every age class.

The conclusions above concerning the relative advantage and disadvantages of scramble and contest competition are based upon a comparison of relative equilibrium levels. If other criteria are used, the conclusions may not be the same. For example, as mentioned above, resilience of equilibrium can also be used as a criterion, greater resilience being interpreted as advantageous. In this case, it turns out (at least for $n > 1$ near 1) that the conclusions above are exactly reversed [230]!

These results show that when making such assertions about relative advantages or disadvantages, properties of the nonlinearity and the methods of comparison are crucial.

Intraspecific predation (cannibalism) can also be studied using hierarchical models under the assumption that victims are younger (smaller) than cannibals. See [30], [100], [104], [114].

3.3.2 Discrete matrix models. An analog of Theorem 3.3.1 for hierarchical age-structured populations can be obtained for a general class of structured matrix models. Consider a projection matrix $P = T + F$ with

$$T = \left[\tau_{ij}\sigma_j\right], \quad F = \left[\varphi_{ij}\beta_j\right], \tag{3.45}$$

where σ_j is the probability that an individual of class j survives one unit of time, τ_{ij} is the fraction of those survivors that moves to class i (hence, $\sum_{i=1}^{m}\tau_{ij} = 1$ for all j), β_j is the number of surviving offspring from an individual in class j, and φ_{ij} is the fraction of the offspring that lies in class i (hence $\sum_{i=1}^{m}\varphi_{ij} = 1$ for all j). Submodels for survival σ_i and fertility β_i are derived under the assumption that these quantities depend on the density y_i of individuals of rank less than i, i.e.,

$$y_i \doteq \begin{cases} 0, \ i = 1, \\ \sum_{j=1}^{i-1} x_j, \quad i \in I[2, m+1], \end{cases}$$

in a population of size $p = y_{m+1}$. This is done in the following way. Let $\beta \in C^0\left(R^2, R_+^1\right)$ be a function such that the per capita birth rate of an individual is $\beta(z,p)$ when the total population size is p and the density of individuals of lower rank is z. Define β_j to be the average

$$\beta_j \doteq \begin{cases} \frac{1}{x_j}\int_{y_j}^{y_j+x_j}\beta(z,p)dz \ \text{ if } x_j \neq 0, \\ \beta(y_j,p) \ \text{ if } x_j = 0\,. \end{cases}$$

For example, if rank is determined by age, the birth rate of individuals in the jth age class ranges from $\beta(y_j,p)$ for the youngest in the class to $\beta(y_{j+1},p) = \beta(y_j + x_j, p)$ for the oldest in the class; β_j is the average for the jth class. Similarly, define

$$\sigma_j \doteq \begin{cases} \frac{1}{x_j}\int_{y_j}^{y_j+x_j}\sigma(z,p)dz \quad \text{if } x_j \neq 0, \\ \sigma(y_j,p) \quad \text{if } x_j = 0, \end{cases}$$

where $\sigma \in C^0(R^2, [0,1])$ is a function such that $\sigma(y,p)$ is the probability an individual will survive one unit of time when the density of younger individuals is y and the total population size is p.

Cumulative sums of the equations

$$x_i(t+1) = \sum_{j=1}^{m} \tau_{ij}\sigma_j x_j(t) + \sum_{j=1}^{m} \varphi_{ij}\beta_j x_j(t)$$

yield

$$y_i(t+1) = f_i\left(y_1(t), \dots, y_m(t), y_{m+1}(t)\right), \quad i \in I[1, m+1],$$

where $f_1 = 0$ and for $i \in I[2, m+1]$

$$\begin{aligned} f_i\left(y_1, \dots, y_m, y_{m+1}\right) \quad &\doteq \quad \sum_{j=1}^{m}\left(\sum_{k=1}^{i-1} \tau_{kj}\right) \int_{y_j}^{y_{j+1}} \sigma(z, y_{m+1})dz \\ &\quad + \sum_{j=1}^{m}\left(\sum_{k=1}^{i-1} \varphi_{kj}\right) \int_{y_j}^{y_{j+1}} \beta(z, y_{m+1})dz. \end{aligned}$$

Since $y_{m+1} = p$ and $\sum_{k=1}^{m} \tau_{kj} = \sum_{k=1}^{m} \varphi_{kj} = 1$, the last equation in this system of difference equations uncouples as a scalar equation for total population size. Thus, letting

$$y = \begin{bmatrix} y_1 \\ y_2 \\ \vdots \\ y_m \end{bmatrix}, \quad f(y,p) = \begin{bmatrix} 0 \\ f_2(y_1, \dots y_{m-1}, p) \\ \vdots \\ f_m(y_1, \dots y_{m-1}, p) \end{bmatrix}$$

we obtain the system of difference equations

$$y(t+1) = f(y(t), p(t)) \tag{3.46}$$

for the cumulative distributions $y_i(t)$, $i \in I[2,m]$, and the uncoupled scalar equation

$$p(t+1) = s(p(t)) + b(p(t)) \tag{3.47}$$

for total population size $p(t) = y_{m+1}(t)$, where $s, b \in C^0\left(R^1, R^1_+\right)$ are functions defined by

$$s(p) \doteq \int_0^p \sigma(z,p)dz, \quad b(p) \doteq \int_0^p \beta(z,p)dz.$$

This is a generalization of the result in [444] (where it is assumed that all newborns lie in the first class).

The dynamics of total population $p(t)$ can be studied by means of the many analytical and graphical techniques available for one-dimensional maps. If the dynamics of $p(t)$ are known, the dynamics of $y(t)$ (and consequently of the original class distribution $x(t)$) can be studied by means of the resulting nonautonomous system (3.46). For example, if $p(t)$ tends to an equilibrium p, then the limiting equation of (3.46) is the autonomous system

$$q(t+1) = f(q(t), p).$$

For the special case of a Leslie matrix it is shown in [444] that if $p(t)$ equilibrates, then $y_i(t)$ and $x_i(t)$ also equilibrate. The following theorem is a discrete analog of Theorem 3.3.1 [444].

THEOREM 3.3.2. *Suppose that the projection matrix $T+F$ described by (3.45) is a Leslie matrix, i.e., $\tau_{ij} = 0$ for all $j \neq i+1$, $\tau_{i,i+1} = 1$ and $\varphi_{ij} = 0$ for all $i \neq 1$ and j, $\varphi_{1j} = 1$. Assume that the functions β, σ and their partial derivatives $\partial_p\beta$, $\partial_p\sigma$ are bounded on R^2_+. Suppose that a solution $p(t) \geq 0$, $t \in I[0, +\infty)$, of (3.47) satisfies $\lim_{t\to+\infty} p(t) = p \geq 0$, and let $x_i(t)$ be the solution of the matrix model whose initial conditions satisfy $\sum_{i=1}^m x_i(0) = p(0)$, $x_i(0) \geq 0$. Then $\lim_{t\to+\infty} y_i(t) = y_i$ and $\lim_{t\to+\infty} x_i(t) = x_i$ exist and are given by the formulas*

$$y_i = \begin{cases} 0, & i = 1, \\ \int_0^p \beta(z,p)dz + \int_0^{y_{i-1}} \sigma(z,p)dz, & i \in I[2,m], \end{cases}$$

$$x_i = \begin{cases} \int_0^p \beta(z,p)dz, & i = 1, \\ \int_{y_{i-1}}^{y_i} \sigma(z,p)dz, & i \in I[2, m-1], \\ \int_{y_{m-1}}^{p} \sigma(z,p)dz, & i = m. \end{cases}$$

Contest versus scramble competition revisited. Consider the comparison of contest and scramble competition in a hierarchical population. Assume that survivorship is a constant $\sigma(z,p) = \sigma_0 > 0$. For scramble competition let $\beta = \beta_0 u\,(Rc_s(p))$, where $Rc_s(p)$ is the per capita resource availability for any individual and $u\,(Rc_s(p))$ is the amount of available resource consumed by an individual. For contest competition let $\beta = \beta_0 u\,(Rc_c(p-z))$, where $Rc_c(p-z)$ is the resource available to an individual in a population of size p with a density z of lower ranking individuals and $u\,(Rc_c(p-z))$ is the amount of resource consumed by an individual. In both cases $\beta_0 > 0$ is the number of offspring produced per unit resource. The resource uptake rate u and the functions c_s and c_c are assumed to satisfy (3.41) and (3.42). The average per capita resource availability for the ith class under scramble competition is $u\,(Rc_s(p))$ and under contest competition is

$$\begin{cases} \frac{1}{x_j} \int_{y_j}^{y_j+x_j} Rc_c(p-z)dz & \text{if } x_j \neq 0, \\ Rc_c(y_j, p) \ \text{ if } x_j = 0\ . \end{cases}$$

We require that the same amount of resource be divided up under both types of competition, so that

$$\sum_{i=1}^m Rc_s(p)x_i = \sum_{i=1}^m \int_{y_j}^{y_j+x_j} Rc_c(p-z)dz$$

and hence

$$c_s(p) = \frac{1}{p}\int_0^p c_c(z)dz.$$

From (3.47) we have the scalar difference equations

$$\text{scramble:} \qquad p(t+1) = \left[\sigma + \beta_0 u\left(R\frac{1}{p(t)}\int_0^{p(t)} c(z)dz\right)\right]p(t),$$

$$\text{contest:} \qquad p(t+1) = \left[\sigma + \beta_0 \frac{1}{p(t)}\int_0^{p(t)} u(Rc(z))dz\right]p(t)$$

for total population sizes in the two cases. Here the subscript on c_c has been dropped. The equilibrium equations are

$$\text{scramble:} \qquad nf_s(p) = 1,$$

$$\text{contest:} \qquad nf_c(p) = 1,$$

where $n = u(R)\beta_0(1-\sigma)^{-1}$ and f_s and f_c are defined by (3.44). These equilibrium equations have the same form as those of the continuous age-structured model considered in section 3.3.1. As a result we can draw the same conclusions concerning the relationship between the equilibrium levels for scramble and contest competition and how this relationship depends on the concavity of u. This discrete matrix hierarchical model is not necessary age structured, however.

3.4 Total population size in age-structured models

In section 3.1.2 an uncoupled limiting equation for total population size is derived for separable McKendrick age-structured equations (3.28). In those types of equations the fertility rate β is independent of population density. In this section we consider some models in which the focus is instead on density dependence in fertility and for which a limiting equation for total population size can be derived [85].

Consider the McKendrick model (2.5) with $a_M = +\infty$ and

$$\delta = \text{ constant} > 0, \quad \beta(p,a) = nb(p,a), \quad b \in C^1\left(R_1^2, R_+^1\right),$$

where the normalization (2.15) holds, i.e.,

$$\int_0^{+\infty} b(0,a)e^{-\delta a}da = 1,$$

so that n is the inherent net reproductive number. If we assume that newborns (age $a=0$) are not reproductive and fertility is bounded, then we have

$$b(p,0) \equiv 0, \quad b(p,a) \le b_0 < +\infty, \quad a,p \in R_+^1.$$

From the McKendrick partial differential equation (2.5(a)) and its initial condition $\rho_0(a) \in L^1(R^1_+, R^1_+)$ we obtain

$$\rho(t,a) = \begin{cases} \rho_0(a-t)e^{-\delta t}, & a \geq t, \\ \rho(t-a,0)e^{-\delta a}, & t > a, \end{cases}$$

where the total birth rate $\rho(t,0)$ is determined by the birth equation in the McKendrick model (2.5(b)), namely,

$$\rho(t,0) = \psi(p(t),t) + n\int_0^t b(p(t),a)\rho(t-a,0)e^{-\delta a}da, \tag{3.48}$$

where

$$\psi(p(t),t) \doteq ne^{-\delta t}\int_t^{+\infty} b(p(t),a)\rho_0(a-t)da.$$

An integration of the partial differential equation (2.5(a)) with respect to a from $a = 0$ to $+\infty$ yields, after an integration by parts, the equation

$$p'(t) = -\delta p(t) + \rho(t,0). \tag{3.49}$$

The equations (3.48)–(3.49) constitute a coupled system of equations for the total birth rate $\rho(t,0)$ and total population size $p(t)$. This system is equivalent to the system consisting of (3.48) and the equation

$$p'(t) = -\delta p(t) + \psi(p(t),t) + n\int_0^t b(p(t),a)\rho(t-a,0)e^{-\delta a}da.$$

This equation can be rewritten

$$p'(t) = -\delta p(t) + \psi(p(t),t) - n\int_0^t b(p(t),a)\frac{\partial}{\partial a}\left[p(t-a)e^{-\delta a}\right]da$$

and simplified, by an integration by parts, to the uncoupled equation

$$p'(t) = -\delta p(t) + \xi(p(t),p(0),t) + n\int_0^t \partial_a b(p(t),a)p(t-a)e^{-\delta a}da,$$

where $\xi(x,y,t) \doteq \psi(x,t) - nb(x,t)ye^{-\delta t}$. Since $\xi(x,y,t)$ tends to 0 (uniformly for x and y on compact intervals) as $t \to +\infty$, we obtain the limiting equation

$$q'(t) = -\delta q(t) + n\int_0^{+\infty} \partial_a b(q(t),a)q(t-a)e^{-\delta a}da \tag{3.50}$$

for total population size [278], [325] to which we now turn our attention.

The integro-differential equation (3.50) has, of course, the extinction equilibrium $q = 0$. Positive equilibria are solutions $q \in R^1_+$ of the equation

$$-\delta + n\int_0^{+\infty} \partial_a b(q,a)e^{-\delta a}da = 0$$

or equivalently of the equation

$$\begin{aligned} &n\nu(q) = 1, \\ &\nu(p) \doteq \textstyle\int_0^{+\infty} b(p,a)e^{-\delta a}da. \end{aligned} \tag{3.51}$$

Stability can be studied formally by linearizing (3.50) at an equilibrium $q \geq 0$ to obtain the linear integro-differential equation

$$y'(t) + cy(t) + \int_0^{+\infty} k(a)y(t-a)da = 0,$$

where

$$k(a) \doteq -n\partial_a b(q,a)e^{-\delta a}, \qquad c \doteq \delta\left(1 - qn\partial_p \nu(q)\right)$$

and investigating solutions $y = e^{zt}$ exponential in time. This results in the characteristic equation

$$f(z) \doteq z + c + \int_0^{+\infty} k(a)e^{-az}da = 0$$

for complex z. If the characteristic function $f(z)$ has no complex roots satisfying $\operatorname{Re} z \geq 0$, then the equilibrium q is stable; if there exists a root with $\operatorname{Re} z > 0$, then the equilibrium is unstable [325].

For the extinction equilibrium $q = 0$ the roots of the characteristic function

$$\begin{aligned} f(z) &= z + \delta - n\int_0^{+\infty} \partial_a b(0,a)e^{-\delta a}e^{-az}da \\ &= (z+\delta)\left(1 - n\int_0^{+\infty} b(0,a)e^{-\delta a}e^{-az}da\right) \end{aligned}$$

are $z = -\delta < 0$ and the solutions of the equation

$$n\int_0^{+\infty} b(0,a)e^{-\delta a}e^{-az}da = 1.$$

If $n < 1$ and $\operatorname{Re} z \geq 0$, the left-hand side is less than 1 in magnitude. Hence the equation has no roots satisfying $\operatorname{Re} z \geq 0$. If $n > 1$, then the left-hand side, as a function of $x = \operatorname{Re} z$, is greater than 1 at $x = 0$, strictly decreases to 0 as $x \to +\infty$, and consequently has a unique positive real root. As a result we obtain the familiar result that the extinction equilibrium loses stability as n is increased through 1.

Let $q > 0$ be a positive equilibrium. By Theorem C.0.1 (see Appendix C) q is unstable if

$$f(0) = -\delta qn\partial_p \nu(q) < 0.$$

Suppose, on the other hand, that $f(0) > 0$. A calculation shows

$$f(ir) = (\delta + ir)\left(1 - n\int_0^{+\infty} b(q,a)e^{-(\delta+ir)a}da\right) + f(0).$$

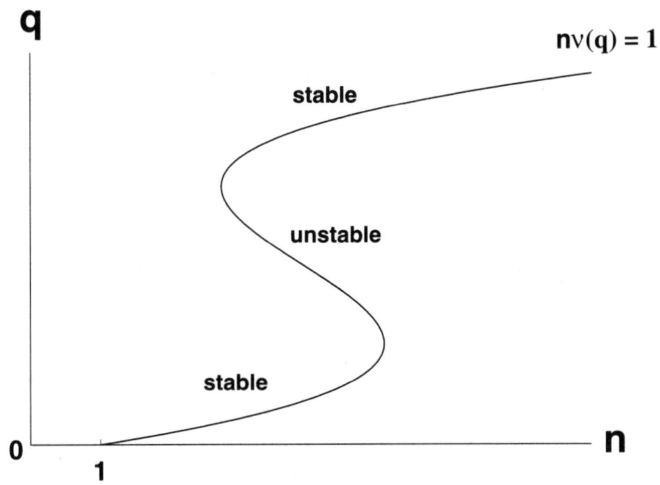

FIG. 3.5. *A plot of the graph defined by* $n\nu(q) = 1$ *gives the equilibrium bifurcation diagram in the* n, q *plane together with stability properties.*

Theorem C.0.1 (see Appendix C) implies that the number of roots of $f(z)$ lying in the right half complex plane is equal to $2m$, where m is the number of times the image of the upper imaginary axis winds around the origin. Consider $f(ir)$ for $r > 0$. Note that the second factor in the product on the right-hand side of $f(ir)$ (which by (3.51) lies in the unit circle centered at $z = 1$) has an argument lying between $-\pi/2$ and $\pi/2$ and that the first factor has an argument lying *strictly* between $-\pi/2$ and $\pi/2$. Thus, the product has an argument lying strictly between $-\pi$ and π, which means that the image $f(ir)$ of the imaginary half axis ir, $r > 0$, does not cross the negative real axis. Thus, $m = 0$ and there are no roots in the right half complex plane. It follows that the equilibrium $q > 0$ is (locally asymptotically) stable.

In summary $q = 0$ loses stability as n increases through 1 and a positive equilibrium $q > 0$ is stable if $\partial_p \nu(q) < 0$ and unstable if $\partial_p \nu(q) > 0$.

This result has a nice interpretation in a bifurcation diagram that plots the equilibrium q against the inherent net reproductive number n. Since equation (3.51) defines the graph of the positive equilibrium, an implicit differentiation shows that those equilibria lying on an increasing branch of the bifurcation curve are stable, while those lying on a decreasing branch of this curve are unstable. See Fig. 3.5. Note in particular that the common, density regulation assumption $\partial_p b(p, a) < 0$ implies that all positive equilibrium are stable. Thus, in such a model, age-dependent fertility (and, in particular, a maturation period delay) cannot destabilize an equilibrium.

APPENDIX A

Stability Theory for Maps

A.1 Linear maps

Consider the linear nonhomogeneous equation

$$x(t+1) = Px(t) + g(t), \tag{A.1}$$

where P is an $m \times m$ matrix of real numbers. For a given sequence $g : I[0,+\infty) \to R^m$ a *forward solution* is defined to be a sequence $x : I[0,+\infty) \to R^m$ that satisfies (A.1) for all $t \in I[0,+\infty)$. For a given sequence $g : I(-\infty,-1] \to R^m$ a *backward solution* is a sequence $x : I(-\infty,0] \to R^m$ that satisfies (A.1) for all $t \in I(-\infty,-1]$.

LEMMA A.1.1. *(Variation of constants formula). For any $g : I[0,+\infty) \to R^m$ the initial value problem*

$$\begin{aligned} x(t+1) &= Px(t) + g(t), \quad t \in I[0,+\infty), \\ x(0) &= x_0 \end{aligned}$$

has a unique forward solution, and this solution is given by the formula

$$x(t) = \begin{cases} P^t x_0 + \sum_{i=0}^{t-1} P^{t-i-1} g(i), & t \in I[1,+\infty), \\ x_0, & t = 0. \end{cases} \tag{A.2}$$

Proof. For $t = 0$ we have $x(1) = Px_0 + g(0) = Px(0) + g(0)$. For $t \in I[1,+\infty)$

$$\begin{aligned} x(t+1) &= P^{t+1} x_0 + \sum_{i=0}^{t} P^{t-i} g(i) \\ &= P\left[P^t x_0 + \sum_{i=0}^{t-1} P^{t-i-1} g(i)\right] + g(t) \\ &= Px(t) + g(t). \end{aligned}$$

Thus, (A.2) satisfies (A.1) for all $t \in I[0,+\infty)$. □

We are interested in forward solutions of (A.1) that are bounded and also those that tend to 0 as $t \to +\infty$. For a sequence $x : I[0,+\infty) \to R^m$ define

$\|x\|_+ \doteq \sup_{t\in I[0,+\infty)} |x(t)|$. The sets

$$\begin{aligned} BS^+ &\doteq \left\{x : I[0,+\infty) \to R^m : \|x\|_+ < +\infty\right\}, \\ BS_0^+ &\doteq \left\{x \in BS^+ : \lim_{t\to+\infty} |x(t)| = 0\right\} \end{aligned}$$

are Banach spaces under the norm $\|\cdot\|_+$. We are also interested in backward solutions of (A.1) that are bounded or tend to 0 as $t \to -\infty$. For a sequence $x : I(-\infty, 0] \to R^m$ define $\|x\|_- \doteq \sup_{t\in I(-\infty,0]} |x(t)|$. The sets

$$\begin{aligned} BS^- &\doteq \left\{x : I(-\infty,0] \to R^m : \|x\|_- < +\infty\right\}, \\ BS_0^- &\doteq \left\{x \in BS^- : \lim_{t\to-\infty} |x(t)| = 0\right\} \end{aligned}$$

are Banach spaces under the norm $\|\cdot\|_-$. Our first goal is to develop modified variation of constants formulas for solutions of (A.1) that lie in these spaces.

Assume that P is hyperbolic, i.e., P has no eigenvalues satisfying $|\lambda| = 1$. Assume that P has Jordan form, i.e.,

$$P = \begin{pmatrix} J_s & 0 \\ 0 & J_u \end{pmatrix}, \tag{A.3}$$

where

$$J_s \text{ is } m_s \times m_s, \quad J_u \text{ is } m_u \times m_u, \quad m = m_s + m_u,$$

each eigenvalue ξ of J_u satisfies $|\xi| > 1$, and each eigenvalue λ of J_s satisfies $|\lambda| < 1$. (This can be done without loss of generality since a linear change of coordinates will put P into Jordan form.) We write $J = P_s + P_u$, where

$$P_s \doteq \begin{pmatrix} J_s & 0 \\ 0 & 0 \end{pmatrix}, \qquad P_u \doteq \begin{pmatrix} 0 & 0 \\ 0 & J_u \end{pmatrix}. \tag{A.4}$$

For any integer $t \in I[1, +\infty)$

$$P_s^t = \begin{pmatrix} J_s^t & 0 \\ 0 & 0 \end{pmatrix}, \qquad P_u^t = \begin{pmatrix} 0 & 0 \\ 0 & J_u^t \end{pmatrix},$$

and $P^t = P_s^t + P_u^t$. For any constant η_s satisfying

$$\max\{|\lambda|\} < \eta_s < 1 \tag{A.5}$$

there exists a constant $c_s > 0$ such that

$$\|P_s^t\| \le c_s \eta_s^t, \quad t \in I[0, +\infty). \tag{A.6}$$

For notational convenience define

$$P_s^{-t} = \begin{pmatrix} J_s^{-t} & 0 \\ 0 & 0 \end{pmatrix}, \qquad P_u^{-t} = \begin{pmatrix} 0 & 0 \\ 0 & J_u^{-t} \end{pmatrix}, \qquad t \in I[1, +\infty),$$

and

$$I_s \doteq \begin{pmatrix} I_{m_s} & 0 \\ 0 & 0 \end{pmatrix}, \qquad I_u \doteq \begin{pmatrix} 0 & 0 \\ 0 & I_{m_u} \end{pmatrix},$$

where I_{m_s} and I_{m_u} are the $m_s \times m_s$ and $m_u \times m_u$ identity matrices, respectively. Note that $P_s^0 = I_s$ and $P_u^0 = I_u$. For any constant η_u satisfying

$$\max\left\{|\xi|^{-1}\right\} < \eta_u < 1 \tag{A.7}$$

there exists a constant $c_u > 0$ such that

$$\|P_u^{-t}\| \le c_u \eta_u^t, \quad t \in I[0, +\infty). \tag{A.8}$$

For $t_1,\ t_2 \in I(-\infty, +\infty)$

$$P_s^{t_1} P_s^{t_2} = P_s^{t_1+t_2},\ P_u^{t_1} P_u^{t_2} = P_u^{t_1+t_2},\ P_s^{t_1} P_u^{t_2} = P_u^{t_1} P_s^{t_2} = 0$$

and for $t \in I(-\infty, +\infty)$

$$PP_s^t = (P_s + P_u)P_s^t = P_s^{t+1},\ PP_u^t = (P_s + P_u)P_u^t = P_u^{t+1}.$$

The following lemma contains a *modified variation of constants formula* for *forward* bounded solutions of (A.1).

LEMMA A.1.2. *Assume that P is hyperbolic and has Jordan form* (A.3). *For $a \in R^m$ and $g \in BS^+$ define*

(A.9)

$$x(t) \doteq \begin{cases} I_s a - \sum_{i=0}^{\infty} P_u^{-1-i} g(i), & t = 0, \\ P_s^t a + \sum_{i=0}^{t-1} P_s^{t-1-i} g(i) - \sum_{i=t}^{\infty} P_u^{t-1-i} g(i), & t \in I[1, +\infty), \end{cases}$$

where P_s and P_u are given by (A.4). *Then*

(a) *x is a forward solution of* (A.1);

(b) *$x \in BS^+$;*

(c) *$g \in BS_0^+$ implies that $x \in BS_0^+$.*

Conversely,

(d) *if $x \in BS^+$ (or BS_0^+) is a forward solution of* (A.1), *then there is an $a \in R^m$ such that x is given by* (A.9).

Proof. (a) First of all, we note that the infinite series in the definition (A.9) of $x(t)$ is convergent. Let $\eta = \max\{\eta_s, \eta_u\}$ and $c = \max\{c_s, c_u\}$. Then $0 < \eta < 1$ and

$$\begin{aligned} \left|\sum_{i=t}^{\infty} P_u^{t-1-i} g(i)\right| &\le c\,\|g\|_+ \sum_{i=t}^{\infty} \eta^{-t+1+i} = c\,\|g\|_+ \frac{\eta}{1-\eta}, \\ \left|\sum_{i=0}^{\infty} P_u^{-1-i} g(i)\right| &\le c\,\|g\|_+ \sum_{i=0}^{\infty} \eta^{1+i} = c\,\|g\|_+ \frac{\eta}{1-\eta}. \end{aligned}$$

For $t = 0$ we have

$$\begin{aligned} Px(0) + g(0) &= P\left\{ I_s a - \sum_{i=0}^{\infty} P_u^{-1-i} g(i) \right\} + g(0) \\ &= P_s a - \sum_{i=0}^{\infty} P_u^{-i} g(i) + I_u g(0) + I_s g(0) \\ &= P_s a - \sum_{i=1}^{\infty} P_u^{-i} g(i) + I_s g(0) \\ &= x(1). \end{aligned}$$

For $t \in I[1, +\infty)$ we have

$$\begin{aligned} Px(t) + g(t) &= P\left\{ P_s^t a + \sum_{i=0}^{t-1} P_s^{t-1-i} g(i) - \sum_{i=t}^{\infty} P_u^{t-1-i} g(i) \right\} + g(t) \\ &= P_s^{t+1} a + \sum_{i=0}^{t-1} P_s^{t-i} g(i) - \sum_{i=t}^{\infty} P_u^{t-i} g(i) + I_s g(t) + I_u g(t) \\ &= P_s^{t+1} a + \sum_{i=0}^{t} P_s^{t-i} g(i) - \sum_{i=t+1}^{\infty} P_u^{t-i} g(i) \\ &= x(t+1). \end{aligned}$$

(b) For $t \in I[1, +\infty)$

$$\begin{aligned} |x(t)| &\le c\eta^t |a| + c\,\|g\|_+ \sum_{i=0}^{t-1} \eta^{t-i-1} + c\,\|g\|_+ \sum_{i=t}^{\infty} \eta^{-t+i+1} \\ &\le c|a| + c\,\|g\|_+ \left[\frac{1}{1-\eta} + \frac{\eta}{1-\eta} \right]. \end{aligned}$$

Thus, $x \in BS^+$.

(c) Let $\varepsilon > 0$ be arbitrary and choose $T = T(\varepsilon) \ge 1$ so that $t \ge T(\varepsilon)$ implies that

$$|g(t)| \le \varepsilon \frac{1}{c} \frac{1-\eta}{1+\eta}.$$

Then for $t \geq T(\varepsilon)+1$

$$\begin{aligned}
|x(t)| &\leq \left\|P_s^t\right\| |a| + \sum_{i=0}^{T(\varepsilon)-1} \left\|P_s^{t-1-i}\right\| |g(i)| \\
&\quad + \sum_{i=T(\varepsilon)}^{t-1} \left\|P_s^{t-1-i}\right\| |g(i)| + \sum_{i=t}^{\infty} \left\|P_u^{t-1-i}\right\| |g(i)| \\
&\leq c\eta^t |a| + c\eta^t \|g\|_+ \sum_{i=0}^{T(\varepsilon)-1} \eta^{-1-i} \\
&\quad + c\varepsilon \frac{1}{c}\frac{1-\eta}{1+\eta} \sum_{i=T(\varepsilon)}^{t-1} \eta^{t-1-i} + c\varepsilon \frac{1}{c}\frac{1-\eta}{1+\eta} \sum_{i=t}^{\infty} \eta^{-t+1+i} \\
&\leq c\eta^t |a| + c\eta^t \|g\|_+ \sum_{i=0}^{T(\varepsilon)-1} \eta^{-1-i} \\
&\quad + \varepsilon \frac{1-\eta}{1+\eta}\frac{1}{1-\eta} + \varepsilon \frac{1-\eta}{1+\eta}\frac{\eta}{1-\eta} \\
&= c\eta^t |a| + c\eta^t \|g\|_+ \sum_{i=0}^{T(\varepsilon)-1} \eta^{-1-i} + \varepsilon
\end{aligned}$$

and therefore $0 \leq \limsup_{t\to+\infty} |x(t)| \leq \varepsilon$. Since ε is arbitrary it follows that $\lim_{t\to+\infty} |x(t)| = 0$ and consequently $x \in BS_0^+$.

(d) From the variation of constants formula (A.2) we have, for $t \in I[1,+\infty)$,

$$\begin{aligned}
x(t) &= \left(P_s^t + P_u^t\right) x_0 + \sum_{i=0}^{t-1} \left(P_s^{t-i-1} + P_u^{t-i-1}\right) g(i) \\
&= P_u^t \left(x_0 + \sum_{i=0}^{\infty} P_u^{-i-1} g(i) \right) \\
&\quad + P_s^t x_0 + \sum_{i=0}^{t-1} P_s^{t-i-1} g(i) - \sum_{i=t}^{\infty} P_u^{t-1-i} g(i).
\end{aligned}$$

The last three terms, on the right-hand side, are bounded for $t \in I[1,+\infty)$, and therefore the first term must be bounded for $t \in I[1,+\infty)$. This implies that the initial condition x_0 must satisfy

$$I_u \left(x_0 + \sum_{i=0}^{\infty} P_u^{-i-1} g(i) \right) = 0.$$

As a result

$$x(t) = P_s^t x_0 + \sum_{i=0}^{t-1} P_s^{t-i-1} g(i) - \sum_{i=t}^{\infty} P_u^{t-1-i} g(i).$$

Let a be any vector in R^m such that

$$I_s a = x_0 + \sum_{i=0}^{\infty} P_u^{-1-i} g(i).$$

Then $P_s^t x_0 = P_s^t a$ and $x(t)$ is given by the formula (A.9). $\square$

The following lemma contains a *modified variation of constants formula* for *backward* bounded solutions.

LEMMA A.1.3. *Assume that P is hyperbolic and has Jordan form* (A.3). *For $a \in R^m$ and $g \in BS^-$ define*

$$x(t) \doteq P_u^t a - \sum_{i=t}^{0} P_u^{t-1-i} g(i) + \sum_{-\infty}^{i=t-1} P_s^{t-1-i} g(i), \qquad t \in I(-\infty, 0],$$

where P_s and P_u are given by (A.4). *Then*

(a) *x is a backward solution of* (A.1);

(b) *$x \in BS^-$*;

(c) *$g \in BS_0^-$ implies that $x \in BS_0^-$.*

Proof. (a) First of all, we note that the infinite series in the definition of x is convergent. Let $\eta = \max\{\eta_s, \eta_u\}$ and $c = \max\{c_s, c_u\}$. Then $\eta < 1$ and

$$\left| \sum_{-\infty}^{i=t-1} P_s^{t-1-i} g(i) \right| \le c \|g\|_- \sum_{-\infty}^{i=t-1} \eta^{t-1-i} = c \|g\|_- \frac{1}{1-\eta}.$$

For $t \in I(-\infty, -1]$

$$\begin{aligned}
Px(t) + g(t) &= P\left\{ P_u^t a - \sum_{i=t}^{0} P_u^{t-1-i} g(i) + \sum_{-\infty}^{i=t-1} P_s^{t-1-i} g(i) \right\} + g(t) \\
&= P_u^{t+1} a - \sum_{i=t}^{0} P_u^{t-i} g(i) + \sum_{-\infty}^{i=t-1} P_s^{t-i} g(i) + I_s g(t) + I_u g(t) \\
&= P_u^{t+1} a - \sum_{i=t+1}^{0} P_u^{t-i} g(i) + \sum_{-\infty}^{i=t} P_s^{t-i} g(i) \\
&= x(t+1).
\end{aligned}$$

Therefore, $x(t)$ is a backward solution of (A.1).

(b) For $t \in I(-\infty, 0]$

$$\begin{aligned}
|x(t)| &\le c\eta^{-t} |a| + c \|g\|_- \sum_{i=t}^{0} \eta^{-t+1+i} + c \|g\|_- \sum_{-\infty}^{i=t-1} \eta^{t-1-i} \\
&\le c|a| + c\|g\|_- \frac{\eta}{1-\eta} + c\|g\|_- \frac{1}{1-\eta}.
\end{aligned}$$

Thus, $x \in BS^-$.

(c) Let $\varepsilon > 0$ be arbitrary and choose $T = T(\varepsilon) \le 0$ so that $t \le T(\varepsilon)$ implies that

$$|g(t)| \le \varepsilon \frac{1}{c} \frac{1-\eta}{1+\eta}.$$

For $t \le T(\varepsilon)$

$$\begin{aligned}
|x(t)| &\le \left\| P_u^t \right\| |a| + \sum_{i=T(\varepsilon)+1}^{0} \left\| P_u^{t-1-i} \right\| |g(i)| \\
&\quad + \sum_{i=t}^{T(\varepsilon)} \left\| P_u^{t-1-i} \right\| |g(i)| + \sum_{-\infty}^{i=t-1} \left\| P_s^{t-1-i} \right\| |g(i)| \\
&\le c\eta^{-t} |a| + c\eta^{-t} \|g\|_- \sum_{i=T(\varepsilon)+1}^{0} \eta^{1+i} \\
&\quad + c\varepsilon \frac{1}{c} \frac{1-\eta}{1+\eta} \sum_{i=t}^{T(\varepsilon)} \eta^{-t+1+i} + c\varepsilon \frac{1}{c} \frac{1-\eta}{1+\eta} \sum_{i=-\infty}^{t-1} \eta^{t-1-i} \\
&\le c\eta^{-t} |a| + c\eta^{-t} \|g\|_- \sum_{i=T(\varepsilon)+1}^{0} \eta^{1+i} + \varepsilon \frac{1-\eta}{1+\eta} \frac{\eta}{1-\eta} + \varepsilon \frac{1-\eta}{1+\eta} \frac{1}{1-\eta} \\
&= c\eta^{-t} |a| + c\eta^{-t} \|g\|_- \sum_{i=T(\varepsilon)+1}^{0} \eta^{1+i} + \varepsilon
\end{aligned}$$

and therefore $0 \le \limsup_{t \to -\infty} |x(t)| \le \varepsilon$. Since $\varepsilon > 0$ is arbitrary, it follows that $\lim_{t \to -\infty} |x(t)| = 0$ and consequently $x \in BS_0^-$. □

A linear change of coordinates in Lemmas A.1.2 and A.1.3 produce formulas for matrices P not in Jordan form.

A.2 Linearization of maps

Consider the nonlinear difference equation

$$x(t+1) = f(x(t)), \quad t \in [0, +\infty). \tag{A.10}$$

For $r > 0$ and $x_e \in R^m$ denote

$$B(r, x_e) \doteq \{x \in R^m : |x - x_e| < r\}.$$

Assume that

$$f(x_e) = x_e , \quad f \in C^2 \left(B(r_0, x_e), R^m \right) \tag{A.11}$$

for some $r_0 > 0$. Write equation (A.10) as

$$x(t+1) = Ax(t) + h(x(t)), \tag{A.12}$$

where $x(t)$ has been replaced by $x(t) - x_e$, $A = J_x f(x_e)$ is the Jacobian of f evaluated at x_e, and $h : B(r_0, 0) \to R^m$ is twice continuously differentiable with $h(0) = 0$ and $J_x h(0) = 0$.

The equilibrium $x_e \in R_+^m$ is called *stable* if for each $\varepsilon > 0$ there exists a $\delta = \delta(\varepsilon) > 0$ such that $|x(0) - x_e| < \delta$ implies that $|x(t) - x_e| < \varepsilon$ for all $t \in I[0, +\infty)$. If x is not stable, it is called *unstable*. The equilibrium x_e is an *attractor* if there exists a $\delta > 0$ such that $|x(0) - x_e| < \delta$ implies that $\lim_{t\to+\infty} |x(t) - x_e| = 0$. If x_e is a stable attractor it is called *(locally) asymptotically stable.*

$A = J_x f(x_e)$ is hyperbolic if no eigenvalue satisfies $|\lambda| = 1$. Denote the *spectral radius* of A by $s(A) \doteq \{\max |\lambda| : \lambda \text{ is an eigenvalue of } A\}$. Theorem A.2.1 below appears in [156, Theorem 4.21] although with the unnecessary assumption that $\lambda \neq 0$ (i.e., that f is locally invertible). This theorem can also be found in [278, Theorem 9.14], where a proof is given using Liapunov functions. We will give a proof based on the variation of constants formula and the following discrete version of Gronwall's inequality.

LEMMA A.2.1. *If $k(i) \geq 0$ and $y(i) \geq 0$, $i = 0, 1, 2, \ldots$, are sequences of real numbers such that $y(0) \leq m$ and*

$$y(t) \leq m + \sum_{i=0}^{t-1} k(i)y(i), \qquad t \in I[1, +\infty),$$

for some constant $m \geq 0$, then

$$y(t) \leq m \prod_{i=0}^{t-1} (1 + k(i)), \quad t \in I[1, +\infty). \tag{A.13}$$

Proof. Let the sequence $z(t) \geq 0$ be defined by $z(0) = m$ and

$$z(t) = m + \sum_{i=0}^{t-1} k(i)z(i), \qquad t \in I[1, +\infty).$$

By induction

$$y(t) \leq z(t), \qquad t \in I[0, +\infty). \tag{A.14}$$

For $t \in I[0, +\infty)$

$$z(t+1) = (1 + k(t))\, z(t),$$

and by induction

$$z(t) = m \prod_{i=0}^{t-1} (1 + k(i)), \qquad t \in I[1, +\infty).$$

From (A.14) we obtain (A.13). $\square$

THEOREM A.2.1. *(Fundamental Theorem of Stability). Assume* (A.11). *If* $s\,(J_x f(x_e)) < 1$, *then the equilibrium* $x = x_e$ *of* (A.10) *is asymptotically stable.*

Proof. We show that $x = 0$ is an asymptotically stable equilibrium of equation (A.12). Let $A \doteq J_x f(x_e)$. It is left as an exercise to show that a sequence solves (A.12) if and only if it solves

$$x(t) = \begin{cases} x_0, & t = 0, \\ A^t x_0 + \sum_{i=0}^{t-1} A^{t-i-1} h(x(i)), & t \in I[1, +\infty). \end{cases}$$

For any η satisfying $s\,(J_x f(x_e)) < \eta < 1$ there is a constant $c > 0$ such that $\|A^t\| \leq c\eta^t$, $t \in I[0, +\infty)$. Thus,

$$|x(t)| \leq c\eta^t\,|x_0| + \sum_{i=0}^{t-1} c\eta^{t-i-1}\,|h(x(i))|\,, \quad t \in I[1, +\infty).$$

Let $c' \doteq \max\{1, c\}$. For $\varepsilon = \frac{1}{2c'}\,(1 - \eta)$ choose $\delta > 0$ so that for all $x \in B(\delta, 0)$ the inequality

$$|h(x)| \leq \varepsilon\,|x|$$

holds. This is possible because h is twice continuously differentiable and $h(0) = 0$ and $J_x h(0) = 0$. For as long as $x(t) \in B(\delta, 0)$, $t \in I[1, +\infty)$, we have

$$|x(t)| \leq c'\eta^t\,|x_0| + \sum_{i=0}^{t-1} c'\eta^{t-i-1}\varepsilon\,|x(i)|$$

or, letting $y(t) = \eta^{-t}\,|x(t)|$,

$$y(t) \leq c'\,|x_0| + \sum_{i=0}^{t-1} c'\eta^{-1}\varepsilon y(i).$$

Using $m = c'\,|x_0|\,,\ k(i) = c'\eta^{-1}\varepsilon$ in Lemma A.2.1 we obtain

$$0 \leq y(t) \leq c'\,|x_0|\,\left(1 + c'\eta^{-1}\varepsilon\right)^t$$

and hence

$$|x(t)| \leq c'\,|x_0|\left(\frac{1}{2}\,(\eta + 1)\right)^t \tag{A.15}$$

for as long as $x(t) \in B(\delta, 0)$, $t \in I[0, +\infty)$. Note that $(\eta + 1)\,/2 < 1$. By choosing $|x_0| < \min\{\delta, \delta/c'\}$, we have (by induction) that $x(t) \in B(\delta, 0)$ and hence (A.15) holds for all $t \in I[0, +\infty)$. It follows from inequality (A.15) that the equilibrium $x = 0$ of (A.12) is both stable and an attractor. □

What happens if $s\,(J_x f(x_e)) > 1$? A local stable manifold theorem when f is invertible can be found in [9], [193], [336]. We prove a local stable manifold theorem that does not require f to be invertible.

Let $x(t, x_0)$ denote the solution of (A.10) starting initially at x_0, i.e., $x(0, x_0) = x_0$. If $f(x)$ is q times continuously differentiable, then an induction argument shows that, for each $t \in I[0, +\infty)$, the solution $x = x(t, x_0)$ of the initial value problem $x(t+1) = f(x(t))$, $x(0) = x_0$, is q times continuously differentiable in x_0.

Let E^s denote the span of the generalized eigenvectors associated with eigenvalues of A satisfying $|\lambda| < 1$. E^s is the stable manifold of the linearization, i.e., of the linear homogeneous system

$$x(t+1) = Ax(t), \quad A = J_x g(x_e). \tag{A.16}$$

Let E^u denote the span of the generalized eigenvectors associated with eigenvalues of A satisfying $|\lambda| > 1$. E^u is the unstable manifold of (A.16). Define

$$\begin{aligned} \Sigma^{\pm}(\delta) &\doteq \left\{x \in BS^{\pm} : \|x\|_{\pm} \leq \delta\right\}, \\ \Sigma_0^{\pm}(\delta) &\doteq \left\{x \in BS_0^{\pm} : \|x\|_{\pm} \leq \delta\right\}. \end{aligned}$$

THEOREM A.2.2. *(Stable Manifold Theorem). Assume* (A.11) *and* $J_x f(x_e)$ *is hyperbolic. In a sufficiently small neighborhood of the equilibrium* x_e*, there is*

(a) *a manifold* W^s *of dimension* $\dim E^s$ *passing through* x_e *tangentially to* E^s *such that* $x_0 \in W^s$ *implies that* $\lim_{t\to+\infty} x(t, x_0) = x_e$*, and*

(b) *a manifold* W^u *of dimension* $\dim E^u$ *passing through* x_e *tangentially to* E^u *such that* $x_0 \in W^u$ *implies the existence of at least one backward solution* $x(t, x_0)$ *such that* $\lim_{t\to-\infty} x(t, x_0) = x_e$.

W^s is called the *(local) stable manifold* of $x = x_e$ and W^u the *(local) unstable manifold* of $x = x_e$.

Proof. Without loss of generality assume that a linear change of variables has been performed so that A has Jordan form (A.3) with P_s and P_u defined by (A.4). Choose any $a \in R^m$. By Lemma A.1.2 a solution $x \in BS^+$ of the equations

$$x(t) = \begin{cases} I_s a - \sum_{i=0}^{\infty} P_u^{-1-i} h(x(i)), & t = 0, \\ P_s^t a + \sum_{i=0}^{t-1} P_s^{t-1-i} h(x(i)) \\ \qquad - \sum_{i=t}^{\infty} P_u^{t-1-i} h(x(i)), & t \in I[1, +\infty) \end{cases} \tag{A.17}$$

is a forward bounded solution of (A.12). For fixed a let $N(x, a)$ denote the operator from BS^+ to BS^+ defined by the right-hand side of (A.17). First we show that N is a contraction map from a ball $\Sigma_0^+(\delta)$ into itself, at least for sufficiently small $|a|$.

From (A.11) it follows that corresponding to any $\varepsilon > 0$ there exists a $\delta = \delta(\varepsilon) > 0$ such that for all $x, y \in \Sigma_0^+(\delta)$

$$|h(x(t)) - h(y(t))| \leq \varepsilon\, |x(t) - y(t)|, \quad t \in I[0, +\infty).$$

Choose $\varepsilon = (1-\eta)(1+\eta)^{-1}/2c'$, where $c' \doteq \max\{1, c\}$. For $x \in \Sigma_0^+(\delta(\varepsilon), 0)$ we have

$$|N(x(t),a)| \le \begin{cases} |a| + c\varepsilon\delta \sum_{i=0}^{\infty} \eta^{1+i}, & t = 0, \\ c\,|a| + c\varepsilon\delta \sum_{i=0}^{t-1} \eta^{t-1-i} \\ \qquad + c\varepsilon\delta \sum_{i=t}^{\infty} \eta^{-t+1+i}, & t \in I[1,+\infty), \end{cases}$$

$$\le \begin{cases} |a| + c\varepsilon\delta \frac{\eta}{1-\eta}, & t = 0, \\ c\,|a| + c\varepsilon\delta \frac{1}{1-\eta} + c\varepsilon\delta\frac{\eta}{1-\eta}, & t \in I[1,+\infty). \end{cases}$$

Therefore, for all $t \in I[0,+\infty)$

$$|N(x(t),a)| \le c'\,|a| + c'\varepsilon\delta\frac{1+\eta}{1-\eta} \le c'\,|a| + \frac{1}{2}\delta.$$

If a is chosen so that

$$|a| \le \frac{1}{2c'}\delta, \tag{A.18}$$

then $\|N(x,a)\|_+ \le \delta$. Since $h(x(t)) \in BS_0^+$, it follows from Lemma A.1.2(c) that $N(x(t),a) \in BS_0^+$. Thus, for $a \in R^m$ satisfying (A.18)

$$N : \Sigma_0^+(\delta(\varepsilon)) \to \Sigma_0^+(\delta(\varepsilon)).$$

Next we show that N is a contraction. For $x, y \in \Sigma_0^+(\delta(\varepsilon))$ the inequalities

$$|N(x(t),a) - N(y(t),a)| \le \begin{cases} \sum_{i=0}^{\infty} c'\eta^{1+i}\varepsilon\,|x(i)-y(i)|, & t = 0, \\ \sum_{i=0}^{t-1} c'\eta^{t-1-i}\varepsilon\,|x(i)-y(i)| \\ \quad + \sum_{i=t}^{\infty} c'\eta^{-t+1+i}\varepsilon\,|x(i)-y(i)|, & t \in I[1,+\infty), \end{cases}$$

$$\le \begin{cases} c'\varepsilon\,\|x-y\|_+ \frac{\eta}{1-\eta}, & t = 0, \\ c'\varepsilon\,\|x-y\|_+ \frac{1}{1-\eta} \\ \quad + c'\varepsilon\,\|x-y\|_+ \frac{\eta}{1-\eta}, & t \in I[1,+\infty), \end{cases}$$

show that

$$\|N(x,a) - N(y,a)\|_+ \le c'\varepsilon\frac{1+\eta}{1-\eta}\,\|x-y\|_+ \le \frac{1}{2}\,\|x-y\|_+ \,.$$

This shows that for all a satisfying (A.18) the operator N is a contraction from $\Sigma_0^+(\delta(\varepsilon))$ into itself. We conclude that for all such a, the equation (A.17) has a unique fixed point $x(t,a) \in \Sigma_0^+(\delta(\varepsilon))$. This fixed point is a forward bounded solution of (A.12) that tends to 0 as $t \to +\infty$.

Finally, we need to describe the set of initial conditions x_0 corresponding to the set of fixed points $x(t) = x(t, x_0)$ found above. From (A.17) x_0 satisfies the equation

$$x_0 = I_s a - \sum_{i=0}^{\infty} P_u^{-1-i} h(x(i, x_0)). \tag{A.19}$$

What kind of solution set do these equations define near $x_0 = 0$? Let

$$x_0 = \begin{bmatrix} x_s \\ x_u \end{bmatrix}, \quad a = \begin{bmatrix} a_s \\ a_u \end{bmatrix}, \quad h(x) = \begin{bmatrix} h_s(x) \\ h_u(x) \end{bmatrix},$$

where

$$I_s x_0 = \begin{bmatrix} x_s \\ 0 \end{bmatrix}, \quad I_u x_0 = \begin{bmatrix} 0 \\ x_u \end{bmatrix}, \quad \text{etc.}$$

If we denote $x(t, x_0) = x(t, x_s, x_u)$, then (A.19) can be written

$$\begin{aligned} x_s &= a_s, \\ x_u &= -\sum_{i=0}^{\infty} P_u^{-1-i} h_u(x(i, x_s, x_u)), \end{aligned}$$

from which it is seen that $x_s = a_s$ is arbitrary and $x_u = x_u(a_s)$ satisfies the equation

$$x_u = -\sum_{i=0}^{\infty} P_u^{-1-i} h_u(x(i, a_s, x_u)). \tag{A.20}$$

By the implicit function theorem this equation has a twice continuously differentiable solution $x_u(a_s)$ satisfying $x_u(0) = 0$ for $a_s \approx 0$. Thus, the manifold of initial conditions x_0 is described parametrically by the equations

$$\begin{aligned} x_s &= a_s, \\ x_u &= x_u(a_s), \; x_u(0) = 0. \end{aligned}$$

From equation (A.20) and the fact that h_u is of second order near $x = 0$ it follows that $\nabla_{a_s} x_u(0) = 0$. This proves part (a).

(b) Choose any $a \in R^m$. By Lemma A.1.3 a solution $x \in BS^-$ of

$$x(t) = P_u^t a - \sum_{i=t}^{0} P_u^{t-1-i} h(x(i)) + \sum_{-\infty}^{i=t-1} P_s^{t-1-i} h(x(i)), \quad t \in I(-\infty, 0],$$

is a backward bounded solution of (A.12). For fixed a, let $N(x, a)$ denote the operator from BS^- to BS^- defined by the right-hand side of this equation. Arguments similar to those in (a) above show that N is a contraction map from a ball $\Sigma_0^-(\delta)$ into itself, at least for sufficiently small $|a|$. The unique fixed point $x = x(t, x_0)$ is a backward solution that tends to 0 as $t \to -\infty$. The initial conditions of this set of solutions are parameterized by

$$\begin{aligned} x_s &= x_s(a_u), \quad x_s(0) = 0, \\ x_u &= a_u, \end{aligned}$$

where $x_s(a_u)$ solves the equation $x_s = \sum_{-\infty}^{i=-1} P_s^{-1-i} h(x(i, x_s, a_u))$. □

What can be said about solutions whose initial points do not lie on either of the two manifolds described in Theorem A.2.2? For flows this question is answered by the classical Hartman/Grobman Theorem. This theorem has an analog for maps that are invertible [193]. For noninvertible maps this classical theorem does not hold, as the following example shows.

Consider the $m = 1$ dimensional map $x(t+1) = x^2(t)$. The Jacobian at the equilibrium $x = 0$ is $A = 0$ and the linearization is $y(t+1) = 0$. Does there exist a homeomorphism $H : R \to R$ such that $H(x(t+1, x_0)) = A^{t+1}H(x_0)$, i.e., such that $H(x^2(t)) = 0$ for all $t \in I[0, +\infty)$? Such a map would have to send all positive x to 0 and therefore could not be one-to-one.

We will replace the Hartman/Grobman Theorem with theorems based on the following lemma (which is a special case of Lemma 1 in [84]).

LEMMA A.2.2. *Suppose that X and Y are Banach spaces and $L : X \to Y$ is a closed linear operator. Suppose that there exists a subspace S of X such that the restriction of L to S, denoted by L_S, is one-one and onto Y. Suppose that $h : X \to Y$, $h(0) = 0$, satisfies $|h(x) - h(y)|_X \le \varepsilon |x - y|_Y$ for all x, $y \in \Sigma(\delta) \doteq \{x \in X : \|x\| < \delta\}$ and some constants ε, $\delta > 0$. If $\varepsilon \left\|L_S^{-1}\right\| < 1$, then there exists a constant $c > 0$ such that there is a one-one bicontinuous map between all solutions of $Lx = 0$ in $\Sigma(c\delta)$ and all solutions of $Lx = h(x)$ in $\Sigma(\delta)$.*

Let $X = Y = BS^+$. Lemma A.2.2 can be applied to the bounded linear operator $L : X \to Y$ defined by

$$Lx \doteq x(t+1) - Ax(t), \quad t \in I[0, +\infty),$$

in the following way. Let $X_1 \subseteq R^m$ be the subspace of initial conditions $x(0)$ for which the linear homogeneous system $x(t+1) = Ax(t)$ has a forward bounded solution. Let X_2 be any supplementary subspace: $R^m = X_1 \oplus X_2$. Define $S \doteq \{x \in BS^+ : x(0) \in X_2\}$. To apply Lemma A.2.2, for an appropriate h, we need only show that the restriction of L to S is one-one and onto Y. For this purpose we assume that

$$\text{(A.21)} \qquad \begin{array}{l} \text{for all } g \in BS^+ \text{ there exists at least} \\ \text{one forward solution } x \in BS^+ \text{ of} \\ x(t+1) = Ax(t) + g(t). \end{array}$$

Under this assumption write $x(0) \in X_1 \oplus X_2$ as $x(0) = x_1 + x_2$, $x_i \in X_i$ and define $x_1(t)$ to be the solution of (A.16) with $x_1(0) = x_1$. By definition of X_1, it follows that $x_1(t) \in BS^+$. Then $z(t) = x(t) - x_1(t)$ lies in BS^+ and satisfies $z(t+1) = Az(t) + g(t)$, $z(0) = x_2 \in X_2$, i.e., $Lz = g$ and $z \in S$. This shows that L_S is onto $Y = BS^+$. In order to show that L_S is also one-one, suppose for $g \in BS^+$ there exist $x, y \in S$ such that $Lx = Ly = g$. Then $L(x-y) = 0$ and the difference $x-y \in S$ is a forward bounded solution of (A.16) with initial condition $x(0) - y(0) \in X_2$. By the definition of X_2 this implies that $x(0) - y(0) = 0$ and hence $x(t) = y(t)$, $t \in I[0, +\infty)$.

We are interested in the case when h is higher order at $x = 0$. In this case the condition on h in Lemma A.2.2 is satisfied for $\varepsilon, \delta > 0$ sufficiently small. By

Lemma A.1.2 condition (A.21) holds if A is hyperbolic. In other words, Lemma A.2.2 applies to (A.10)–(A.11) when $A = J_x f(x_e)$ is hyperbolic.

Similar (virtually identical) arguments apply with BS^+ replaced by BS_0^+.

THEOREM A.2.3. *Assume that* (A.11) *holds and* $A = J_x f(x_e)$ *is hyperbolic. There exist constants* c, $\delta > 0$ *such that the following hold.*

(a) *There is a one-one bicontinuous map between forward solutions of* (A.10) *lying in* $\Sigma^+(\delta)$ *and forward solutions of its linearization* (A.16) *lying in* $\Sigma^+(c\delta)$.

(b) *There is a one-one bicontinuous map between forward solutions of* (A.10) *lying in* $\Sigma_0^+(\delta)$ *and forward solutions of its linearization* (A.16) *lying in* $\Sigma_0^+(c\delta)$.

Similar arguments can be made for the bounded linear operator $L : X \to Y$ defined by

$$Lx \doteq \begin{cases} x(0), & t = 0, \\ x(t+1) - Ax(t), & t \in I(-\infty, -1], \end{cases}$$

with $X = Y = BS^-$ or BS_0^-. In this case $X_1 \subseteq R^m$ is the subspace of initial conditions $x(0)$ for which the linear homogeneous system (A.16) has a backward bounded solution, X_2 is any supplementary subspace, and $S \doteq \{x \in BS^- : x(0) \in X_2\}$.

THEOREM A.2.4. *Assume that* (A.11) *holds and* $A = J_x f(x_e)$ *is hyperbolic. There exist constants* c, $\delta > 0$ *such that the following hold.*

(a) *There is a one-one bicontinuous map between backward bounded solutions of* (A.10) *lying in* $\Sigma^-(\delta)$ *and backward solutions of its linearization* (A.16) *lying in* $\Sigma^-(c\delta)$.

(b) *There is a one-one bicontinuous map between backward solutions of* (A.10) *lying in* $\Sigma_0^-(\delta)$ *and backward solutions of its linearization* (A.16) *lying in* $\Sigma_0^-(c\delta)$.

APPENDIX B

Bifurcation Theorems

B.1 A global bifurcation theorem

The set of *characteristic values* of an $m \times m$ matrix L is the set of reciprocals of the nonzero eigenvalues of L. If λ_0 is a characteristic value of L, then there exist nonzero *left and right characteristic vectors* q and v such that $v = \lambda_0 L v$ and $q^\tau = \lambda_0 q^\tau L$. Necessarily a characteristic value is nonzero. An *algebraically simple* characteristic value is a simple root of the characteristic polynomial $\det(I - \lambda L)$. A characteristic value λ_0 is *geometrically simple* if the left and right null spaces of $I - \lambda_0 L$ have dimension 1 and are therefore spanned by single characteristic vectors q and v, respectively. Denote

$$\begin{aligned} V &\doteq \{y \in R^m : (I - \lambda_0 L)\, y = 0\} = \text{Span}\{v\}, \\ Q &\doteq \{y \in R^m : y^\tau (I - \lambda_0 L) = 0\} = \text{Span}\{q\}. \end{aligned} \tag{B.1}$$

An algebraically simple characteristic value is geometrically simple, but the converse is not necessarily true.

Consider the algebraic equation

$$x = \lambda L x + h(\lambda, x), \qquad (\lambda, x) \in R^1 \times \Omega, \tag{B.2}$$

where

$$\begin{aligned} &h : R^1 \times \Omega \to R^m \text{ is continuous and} \\ &|h(\lambda, x)| = O\left(|x|^2\right) \text{ near } x = 0 \\ &\text{uniformly on compact } \lambda \text{ intervals.} \end{aligned} \tag{B.3}$$

Here Ω is an open neighborhood of R_+^m (the closed positive cone). A pair $(\lambda, x) \in R^1 \times \Omega$ that satisfies (B.2) is called a *solution pair.* Any *trivial pair* $(\lambda, 0)$ is a solution pair for all $\lambda \in R^1$. A solution pair (λ, x) is called *nontrivial* if $x \neq 0$. The closure of the set of nontrivial solution pairs is denoted by S. A solution pair is *nonnegative (positive)* if $x \geq 0$ ($x > 0$). A *continuum* of pairs $(\lambda, x) \in R^1 \times R^m$ is a closed and connected set in $R^1 \times R^m$. Let ∂R_+^m denote the boundary of R_+^m, i.e., the set of nonnegative vectors $x \geq 0$ that are not positive. A pair $(\lambda, x) \in R^1 \times \partial R_+^m$ is called a *boundary pair.*

We are interested in the existence of nonnegative nontrivial (and also of positive) solution pairs of (B.2). Specifically, we will look for a continuum in S

that contains a trivial solution pair $(\lambda_0, 0)$ for a critical value λ_0 (i.e., that "bifurcates" from $(\lambda_0, 0)$).

First of all, we observe that the only candidates for such a critical bifurcation value of λ are the characteristic values of the matrix L. To see this suppose there exists a sequence (λ_i, x_i) of nontrivial pairs such that $\lim_{i\to+\infty}(\lambda_i, x_i) = (\lambda_0, 0)$. Define the unit vectors $y_i \doteq x_i / |x_i|$. The unit sphere in R^m is compact so we can assume (by extracting a subsequence if necessary) that $\lim_{i\to+\infty} y_i = y_\infty$, $|y_\infty| = 1$. Since (B.3) implies that $\lim_{i\to+\infty} h(\lambda_i, x_i)/|x_i| = 0$, we find from (B.2), after dividing by $|x_i|$ and passing to the limit $i \to +\infty$, that $y_\infty = \lambda_0 L y_\infty$. Thus, λ_0 is a characteristic value of L.

Is it a sufficient condition for a bifurcation to occur that λ_0 is a characteristic value of L? The answer is in general "no," as the example

$$x_1 = \lambda x_1 + x_2^3,$$
$$x_2 = \lambda x_2 - x_1^3$$

shows. In this example $\lambda_0 = 1$ is the only characteristic value of L (which is the 2×2 identity matrix). For any λ, $x_1 = 0$ implies that $x_2 = 0$ and vice versa. Thus, for a nontrivial solution pair (λ, x) we must have $x_1 \neq 0$ and $x_2 \neq 0$ and therefore $1 - \lambda = x_2^3 x_1^{-1} = -x_1^3 x_2^{-1}$, which leads to the contradiction $x_1^4 + x_2^4 = 0$. This example has no nontrivial solution pairs. Note that $\lambda_0 = 1$ is not a geometrically simple characteristic value.

We will use theorems from [354] to obtain the existence of nontrivial solutions that bifurcate from $(\lambda_0, 0)$ when λ_0 is a geometrically simple characteristic value of L. (In fact, theorems from [354] more generally allow for λ_0 to be of add geometric multiplicity.) Define the continuous function $a : R^m \to R_+^m$ by

$$a(x) = \begin{pmatrix} |x_1| \\ \vdots \\ |x_m| \end{pmatrix}.$$

Since $a(x) = x$ for $x \geq 0$, the set of nonnegative (positive) solution pairs of the equation (B.2) is identical to that of the equation

$$x = \lambda L x + g(\lambda, x), \quad g(\lambda, x) = h(\lambda, a(x)).$$

The advantage of working with this equation is that it is globally defined in x, i.e., $\Omega = R^m$. The function g satisfies (B.3). If we assume that λ_0 has a positive characteristic vector, then Theorem 1.25 of [354] implies that S, the closure of the set of nontrivial solution pairs of this equation, contains a continuum C that "bifurcates" from $(\lambda_0, 0)$ (i.e., $(\lambda_0, 0) \in C$) such that in some open neighborhood of $(\lambda_0, 0)$ the solution pairs from $C/\{(\lambda_0, 0)\}$ are positive. These positive solution pairs are then positive solution pairs of the original equation (B.2). Thus, we have a local bifurcation result for positive solution pairs. We turn now to the problem of determining the global extent of these positive solution pairs.

Theorem 1.40 in [354] implies that the (locally positive) continuum C satisfies one of two alternatives: C is either (1) unbounded in $R^1 \times R^m$ or (2) it contains a trivial solution pair $(\lambda_1, 0)$ for which $\lambda_1 \neq \lambda_0$ is another characteristic value of L. (This theorem also implies the existence of another continuum satisfying these same alternatives which is locally negative. The theorem guarantees that these two continua are globally distinct, however, i.e., they do not meet except at the point $(\lambda_0, 0)$.)

We are interested in only nonnegative (and positive) solution pairs from the continuum C. Let C have the topology inherited from $R^1 \times R^m$. The subset $S^+ \doteq \{(\lambda, x) \in C : x > 0\}$ of positive solution pairs from C is open. Let S_0^+ be the maximal open subset of S^+ whose closure C^+ is connected and contains $(\lambda_0, 0)$. The set S_0^+ is nonempty since the continuum C is positive in a neighborhood of $(\lambda_0, 0)$. By considering the two alternatives (1) and (2) above we conclude the following about the continuum C^+: either $C^+ / \{(\lambda_0, 0)\}$ is unbounded and contains only positive solution pairs or C^+ contains a boundary point $(\lambda^*, x^*) \neq (\lambda_0, 0)$. In the latter case we can say more. If $x^* = 0$, then, as we showed above, λ_1 is a characteristic value of L and it has a nonnegative characteristic vector (since $y_i \geq 0$ implies that $y_\infty \geq 0$).

We summarize these results in the following theorem.

THEOREM B.1.1. *Suppose that* (B.3) *holds and L has a geometrically simple characteristic value λ_0 which has a positive right characteristic vector $v > 0$. Then there exists a continuum C^+ of nonnegative solution pairs of* (B.2) *containing $(\lambda_0, 0)$ for which $C^+ / \{R^1 \times \partial R_+^m\}$ is nonempty and consists of positive solution pairs. The following alternatives hold: either*

(i) $C_0^+ \doteq C^+ / \{(\lambda_0, 0)\}$ *is unbounded in $R^1 \times R_+^m$ and contains only positive solution pairs, or*

(ii) C^+ *contains a boundary pair $(\lambda^*, x^*) \neq (\lambda_0, 0)$.*

In case (ii) *if $x^* = 0$, then $\lambda^* \neq \lambda_0$ is another characteristic value of L which has a nonnegative characteristic vector.*

This theorem could be succinctly described by stating that, under the assumed conditions, there exists a "positive branch" of solutions of (B.2) that bifurcates from $(\lambda_0, 0)$ and "connects" to the boundary ∂R_+^m of the positive cone R_+^m (∞ being considered as on the boundary).

B.2 Local parameterization

In this section we consider a parametric representation of the bifurcating branch of nontrivial solutions guaranteed by Theorem B.1.1. This representation is sometimes called the *Liapunov/Schmidt expansion* or *parameterization* of the branch.

Consider first the linear nonhomogeneous algebraic equation $x = Lx + f$ for an unknown $x \in R^m$, with $f \in R^m$ and the $m \times m$ matrix L given. There exists a unique solution $x = (I - L)^{-1} f$ if and only if $I - L$ is nonsingular. Suppose, however, that $I - L$ is singular and that its null spaces (as in (B.1) with $\lambda_0 = 1$) have dimension 1. If $x = Lx + f$ has a solution, then $q^\tau x = q^\tau Lx + q^\tau f$ implies that $q^\tau (I - L) x = q^\tau f$ and, since $q \in Q$, that $q^\tau f = 0$. Conversely, if $q^\tau f = 0$, then the general solution has the form $x = \varepsilon v + z$, where $\varepsilon \in R^1$ is an arbitrary scalar and z is any particular solution. Out of this general solution we can extract a unique solution that satisfies $v^\tau x = 0$ (by choosing $\varepsilon = -v^\tau z / v^\tau v$). Let $V^\perp$ and $Q^\perp$ denote the spaces orthogonal to V and Q, respectively, i.e.,

$$\begin{aligned} V^\perp &= \{y \in R^m : v^\tau y = 0\}, \\ Q^\perp &= \{y \in R^m : q^\tau y = 0\}. \end{aligned}$$

Then $R^m = Q \oplus Q^\perp = V \oplus V^\perp$. For each $f \in Q^\perp$ the equation $x = Lx + f$ has a unique solution $x \in V^\perp$. This correspondence defines a linear transformation of $Q^\perp$ into $V^\perp$. Let G be the matrix associated with this transformation so that Gf is the unique solution lying in $V^\perp$. The general solution of $x = Lx + f$ has the form $x = \varepsilon v + Gf$, where $\varepsilon \in R^1$ is arbitrary.

Consider the nonlinear algebraic equation

$$x = Lx + f(x) \tag{B.4}$$

when $I - L$ is singular and its null space is one dimensional. Suppose that $x \in R^m$ is a solution. Since $R^m = V \oplus V^\perp$ we can write $x = \varepsilon v + z$, $\varepsilon \in R^1$, $z \in V^\perp$. It must be the case that $f(x) \in Q^\perp$, i.e., $q^\tau f(x) = 0$, and therefore that $z = Gf(x)$. Thus, ε and $z \in V^\perp$ satisfy the pair of equations

$$\begin{aligned} 0 &= q^\tau f(\varepsilon v + z), \\ z &= Gf(\varepsilon v + z). \end{aligned} \tag{B.5}$$

Conversely, if $\varepsilon \in R^1$ and $z \in V^\perp$ satisfy these two equations, then $x = \varepsilon v + z$ satisfies the equation (B.4). In this sense equation (B.4) is equivalent to the pair of equations (B.5).

Consider the equation with parameter

$$x = \lambda L x + h(\lambda, x), \quad h(\lambda, 0) = 0. \tag{B.6}$$

The problem is to find nonzero solution pairs $(\lambda, x) = (\lambda, \varepsilon v + z) \in R^1 \times V \oplus V^\perp$ for λ near a characteristic value λ_0 of L, i.e., for a value such that $I - \lambda_0 L$ is singular. Suppose that λ_0 is a geometrically simple characteristic value of L with left and right characteristic vectors q and v. Define $\eta = \lambda - \lambda_0$, and write

$$x = \lambda_0 L x + [\eta L x + h(\lambda_0 + \eta, x)].$$

With $f(x)$ taken to be $\eta L x + h(\lambda_0 + \eta, x)$ the equivalent pair (B.5) can be written

$$\begin{aligned} &\text{(a)} \quad 0 = \eta \lambda_0^{-1} \varepsilon q^\tau v + \eta q^\tau L z + q^\tau h(\lambda_0 + \eta, \varepsilon v + z), \\ &\text{(b)} \quad 0 = z - \eta \lambda_0^{-1} \varepsilon G v - \eta G L z - G h(\lambda_0 + \eta, \varepsilon v + z). \end{aligned} \tag{B.7}$$

We view this as a system to be solved for $\eta \in R^1$ and $z \in V^\perp$ as functions of $\varepsilon \approx 0$. Using the implicit function theorem on the second equation (B.7(b)), we can locally solve uniquely for

$$z = z(\varepsilon, \eta), \quad z(0,0) = 0,$$

where z is just as smooth in its arguments as h is in its arguments. Assume that

(B.8) $h : R^1 \times R^m \to R^m$ is $k+1$ times continuously differentiable near $(\lambda, x) = (\lambda_0, 0)$ for some $k \geq 1$ and $|h(\lambda, x)| = O\left(|x|^2\right)$ near $x = 0$ uniformly on compact λ intervals.

Then $z(\varepsilon, \eta)$ is $k+1$ times continuously differentiable near $\varepsilon = \eta = 0$. We can use the implicit function theorem because the right-hand side of the second equation (B.7(b)) vanishes when $z = 0$, $\eta = \varepsilon = 0$ and has an invertible Jacobian with respect to z at this point (equal, namely, to the identity). Notice that if $\varepsilon = 0$ in equation (B.7), then $z = 0$ is a solution. By the uniqueness of the solution obtained from the implicit function theorem it follows that $z(0, \eta) = 0$ for all small η and we can "factor an ε out of z"; i.e., we can write $z(\varepsilon, \eta) = \varepsilon\zeta(\varepsilon, \eta)$, where $\zeta(\varepsilon, \eta)$ is k times continuously differentiable near $\varepsilon = \eta = 0$. Also, from equation (B.7(b)) it follows that $\zeta(0, 0) = 0$. With equation (B.7(b)) solved the equivalent system (B.7) reduces, in a neighborhood of $(\lambda_0, 0)$, to the single scalar equation obtained from (B.7(a)), i.e.,

$$0 = \eta\lambda_0^{-1} q^\tau v + \eta q^\tau L\zeta(\varepsilon, \eta) + q^\tau \widehat{h}(\varepsilon, \eta),$$

where $\widehat{h}(\varepsilon, \eta) \doteq \varepsilon^{-1} h\left(\lambda_0 + \eta, \varepsilon v + \varepsilon\zeta(\varepsilon, \eta)\right)$ is k times continuously differentiable near $\varepsilon = \eta = 0$ and $\widehat{h}(0, \eta) = 0$. By the implicit function theorem this equation can be locally solved for a (unique) C^k solution $\eta = \eta(\varepsilon)$, $\eta(0) = 0$ provided $q^\tau v \neq 0$ (since the right-hand side of the equation vanishes at $\varepsilon = \nu = 0$ and has a derivative with respect to η equal to $\lambda_0^{-1} q^\tau v$ at this point). Using $z(\varepsilon) = \varepsilon\zeta(\varepsilon, \eta(\varepsilon))$ we obtain a C^k solution pair $\lambda = \lambda_0 + \eta(\varepsilon)$, $x = \varepsilon v + z(\varepsilon)$ of (B.6) for ε near 0.

THEOREM B.2.1. *Suppose that* (B.8) *holds and L has a geometrically simple characteristic value λ_0 which has left and right characteristic vectors q and v. Assume that $v > 0$ and $q^\tau v \neq 0$. Then in a neighborhood of $(\lambda, x) = (\lambda_0, 0)$ the branch of nontrivial solutions of equation* (B.6) *guaranteed by Theorem* B.1.1 *can be written, for $\varepsilon \in (-\varepsilon_0, \varepsilon_0)$, $\varepsilon_0 > 0$, in the form*

$$\lambda = \lambda_0 + \eta(\varepsilon), \qquad x = \varepsilon v + z(\varepsilon),$$

where $\eta \in C^k((-\varepsilon_0, \varepsilon_0), R^1)$, $z \in C^k((-\varepsilon_0, \varepsilon_0), V^\perp)$ and $\eta(0) = 0$, $z(0) = z'(0) = 0$.

Approximations of the bifurcating branch can be obtained by a calculation of $\eta'(0)$ and $z''(0)$. These quantities can be found by differentiating the equation $x = \lambda Lx + h(\lambda, x)$ twice with respect to ε and evaluating the result at $\varepsilon = 0$. This yields

$$z''(0) = \lambda_0 L z''(0) + 2\eta'(0)Lv + h(v, v).$$

Here $h(v, v)$ is the vector whose i^{th} component is

$$v^\tau H_i v = \sum_{j=1}^{m} v_j \sum_{k=1}^{m} v_k \partial_k \partial_j h_i(\lambda_0, 0),$$

where $H_i = [\partial_k \partial_j h_i(\lambda_0, 0)]$ is the $m \times m$ Hessian of h_i with respect to x evaluated at $(\lambda, x) = (\lambda_0, 0)$. Since λ_0 is a characteristic value of L, the necessary condition for the solvability of this equation for $z'(0)$ is $q^\tau \left(2\eta'(0)Lv + h(v, v)\right) = 0$, which yields the formula

$$\eta'(0) = -\frac{\lambda_0}{2} \frac{q^\tau h(v, v)}{q^\tau v}.$$

The solution for $z''(0)$ is then

$$z''(0) = G\left(h(v, v) - \frac{q^\tau h(v, v)}{q^\tau v} v\right).$$

Thus, if $k \geq 2$, we have the parameterization

$$\begin{aligned} \lambda &= \lambda_0 - \frac{\lambda_0}{2} \frac{q^\tau h(v, v)}{q^\tau v} \varepsilon + O\left(\varepsilon^2\right), \\ z &= \varepsilon v + \frac{1}{2} G\left(h(v, v) - \frac{q^\tau h(v, v)}{q^\tau v} v\right) \varepsilon^2 + O\left(\varepsilon^2\right). \end{aligned}$$

APPENDIX C

Miscellaneous Proofs

Proof of Theorem 1.1.2(b). Part (b) of Theorem 1.1.2 follows immediately from the following lemma.

LEMMA C.0.1. *Suppose that $q(t)$ satisfies the scalar equation $q(t+1) = c(t)q(t)$, $t \in I[0,+\infty)$, where $\lim_{t\to+\infty} c(t) = c_\infty$.*

(a) *If $|c_\infty| < 1$, then $\lim_{t\to+\infty} |q(t)| = 0$ (exponentially).*

(b) *Suppose that $|c_\infty| > 1$. If $c(t) \neq 0$ for all $t \in I[0,+\infty)$, then $\lim_{t\to+\infty} |q(t)| = +\infty$ (exponentially).*

Proof. (a) Pick $\varepsilon > 0$ such that $|c_\infty| + \varepsilon < 1$. There exists a $T = T(\varepsilon) \geq 0$ such that $t \in I[T,+\infty)$ implies that $|c(t)| \leq |c_\infty| + \varepsilon$. For $t \in I[T,+\infty)$, $q(t+1) = q(T)\prod_{i=T}^{i=t} c(i)$ and $|q(t+1)| \leq |q(T)|\left(|c_\infty| + \varepsilon\right)^{t-T+1}$. Hence $\lim_{t\to+\infty} |q(t)| = 0$.

(b) Pick $\varepsilon > 0$ such that $|c_\infty| - \varepsilon > 1$. There exists a $T = T(\varepsilon) \geq 0$ such that $t \in I[T,+\infty)$ implies that $|c(t)| \geq |c_\infty| - \varepsilon$. For $t \in I[T,+\infty)$, $q(t+1) = q(T)\prod_{i=T}^{i=t} c(i)$ and $|q(t+1)| \geq |q(T)|\left(|c_\infty| - \varepsilon\right)^{t-T+1}$. Hence $\lim_{t\to+\infty} |q(t)| = +\infty$.

Proof of Theorem 1.2.1(b). We show that the origin 0 is (a) an isolated invariant set in R_+^m and (b) equal to its own stable set. It follows from Theorem 4.1 in [235] that the origin is uniformly persistent with respect to the origin.

By definition the set $\{0\}$ is isolated if there exists a closed neighborhood U of $\{0\}$ in R_+^m such that $\{0\}$ is the largest invariant set in U. Let M' denote the largest invariant set in U and choose $x(0) \in M'$. Then $x(0) \geq 0$ and the forward orbit of $x(0)$ remains in M' and hence is bounded. Thus $x(0)$ lies on the stable manifold of 0. However, according to the remarks prior to the statement of Theorem 1.2.1, the local stable manifold intersects the closed positive cone at the origin only when $r > 1$. Thus, $x(0) = 0$ and $M' = \{0\}$.

(b) By the Theorems A.2.2, A.2.3, and A.2.4 there is no point $x(0) \in U/\{0\}$ such that $\lim_{t\to+\infty} |x(t)| = 0$. Thus, the stable set of $\{0\}$ is itself.

Proof of Lemma 1.2.2. For small $\varepsilon > 0$ the Jacobian $J_x(\lambda,x) = A + \lambda B + J_x r(\lambda,x)$ evaluated at the equilibrium (1.31) can be written $J_x(\lambda,x) = J_0 + J_1\varepsilon + O(\varepsilon^2)$. From

$$J_x r(\lambda,x) = \left[\sum_{k=1}^{m} \partial_k \partial_j r_i(\lambda_0,0) y_k\right] \varepsilon + O\left(\varepsilon^2\right)$$

we obtain

$$J_0 = A + \lambda_0 I, \quad J_1 = \lambda_1 B + \left[\sum_{k=1}^{m} \partial_k \partial_j r_i(\lambda_0, 0) y_k \right].$$

The dominant eigenvalue and eigenvector of the Jacobian $J_x(\lambda, x)$ also have ε expansions

$$e = v + e_1 \varepsilon + O\left(\varepsilon^2\right) \quad \text{and} \quad \zeta = 1 + \zeta'(0)\varepsilon + O\left(\varepsilon^2\right).$$

Placing these expressions into $J_x(\lambda, x)e = \zeta e$ we obtain, to first order in ε, the equation

$$J_0 e_1 = e_1 + \zeta'(0)v - J_1 v.$$

Necessary for the solution of this linear algebraic system for e_1 is the orthogonality condition $w^\tau(\zeta'(0)v - J_1 v) = 0$, which yields

$$\zeta'(0) = \frac{w^\tau J_1 v}{w^\tau v}.$$

The numerator is

$$w^\tau J_1 y = w^\tau B v \lambda_1 + w^\tau \left(\sum_{k=1}^{m} \partial_k \partial_j r_i(\lambda_0, 0) y_k \right) v = w^\tau B v \lambda_1 + w^\tau h(v, v).$$

Using the formula in (1.31) for λ_1, we obtain

$$w^\tau J_1 v = w^\tau B v \lambda_1 + 2\left(-w^\tau B v \lambda_1\right) = -w^\tau B v \lambda_1$$

and hence

$$\zeta'(0) = -\frac{w^\tau B v}{w^\tau v}\lambda_1.$$

Proof of Theorem 3.1.3. Define $\rho(t) \doteq a(t)/b(t)$. Then from equation (3.9)

$$\varphi(t+1) = \frac{(\rho(t)I + L)\varphi(t)}{w^\tau(\rho(t)I + L)\varphi(t)}, \quad t \in I[0, +\infty),$$

follows. It is not difficult to show that the solution of this equation is $\varphi(t) = \psi(t)/w^\tau\psi(t)$, where ψ is the solution of the equation

$$\psi(t+1) = (\rho(t)I + L)\psi(t), \quad \psi(0) = \varphi(0).$$

For simplicity we give a proof when L has a basis of eigenvectors $v_1 = v$, $v_2, \ldots, v_m$. For the more general case see [76], [289]. Write $\varphi(0) = \sum_{i=1}^{m} c_i v_i$. The matrix $\rho(t)I + L$ has eigenvalues $\lambda_i + \rho(t)$ and eigenvectors v_i. From the solution

$$\psi(t) = \sum_{i=1}^{m} c_i \prod_{k=0}^{t-1} (\lambda_i + \rho(k)) v_i$$

we have

$$\varphi(t) = \frac{\sum_{i=1}^{m} c_i \prod_{k=0}^{t-1} (\lambda_i + \rho(k))\, v_i}{\sum_{i=1}^{m} c_i \prod_{k=0}^{t-1} (\lambda_i + \rho(k))\, w^\tau v_i}.$$

Divide numerator and denominator by $\prod_{k=0}^{t-1} (\lambda_1 + \rho(k))$, and define

$$\Pi_i(t) \doteq \prod_{k=0}^{t-1} \frac{\lambda_i + \rho(k)}{\lambda_1 + \rho(k)}.$$

Then

$$\varphi(t) = \frac{c_1 v + \sum_{i=2}^{m} c_i \Pi_i(t) v_i}{c_1 w^\tau v + \sum_{i=2}^{m} c_i \Pi_i(t) w^\tau v_i}, \qquad c_1 = w^\tau \varphi(0) > 0.$$

By the following lemma, $\lim_{t\to+\infty} \Pi_i(t) = 0$ for $i \in I[2,m]$ and limit (3.10) follows.

LEMMA C.0.2. *If* $\lambda_1 > |\lambda_i|$ *for* $i \in I[2,m]$ *and* $0 \le \rho(t) \le \rho_0 < +\infty$ *for* $t \in I[0,+\infty)$, *then* $\lim_{t\to+\infty} \Pi_i(t) = 0$ *for* $i \in I[2,m]$.

Proof. The ratio $|\lambda_i + x| \,/\, (\lambda_1 + x)$ is continuous for $x > 0$. Moreover, the assumption on λ_1 implies that this ratio is less than 1 and therefore it is bounded away from 1 on bounded x intervals. Thus, $0 < |\lambda_i + x| \,/\, (\lambda_1 + x) < m_i$ for some $m_i < 1$ for all $x \in [0, \rho_0]$. It follows that $|\Pi_i(t)| < m_i^t$.

Roots of characteristic polynomials. The following theorem shows that under certain conditions the number of roots of $f(z) \doteq z + c + K(z)$ lying in the right half complex plane is related to the number of times the image $f(ir)$, $r > 0$, of the upper imaginary axis "winds around the origin."

THEOREM C.0.1. (see [85]). *Assume that* $k_1 \doteq \int_0^{+\infty} |k(a)|\, da < +\infty$, *and consider the analytic function* $f(z) \doteq z + c + K(z)$, *where* $K(z) \doteq \int_0^{+\infty} k(a) e^{-az} da$ *is the Laplace transform of* $k(a)$.

(a) *If* $f(0) < 0$, *then* $f(z)$ *has a positive real root.*

(b) *Assume that* $f(0) > 0$ *and* $\int_0^{+\infty} a\, |k(a)|\, da < +\infty$. *Assume that* $f(z)$ *has no purely imaginary roots. Then* $\arg f(+i\infty) = \frac{\pi}{2} - 2m\pi$ *for some integer* $m \in I(-\infty, +\infty)$ *and the number of roots of* $f(z)$ *lying in the right half complex plane is* $2m$.

Proof. (a) For real $z = x$, $\lim_{x\to+\infty} f(x) = +\infty$ and the result follows by the intermediate value theorem.

(b) By the argument principle the number of roots lying inside the semidisk $|z| \le r$, $\operatorname{Re} z \ge 0$, is

$$\nu(r) = \frac{1}{2\pi i} \oint_{\partial(r)} \frac{f'(z)}{f(z)} dz,$$

where $\partial(r)$ is the boundary of the semidisk. Consider first the integral over the semicircular part $|z| = r$, $\operatorname{Re} z \ge 0$, of the boundary $\partial(r)$, which we denote by

$\partial_1(r)$. Consider the difference

$$\frac{f'(z)}{f(z)} - \frac{1}{z} = \frac{K'(z)}{f(z)} - \frac{c+K(z)}{zf(z)}.$$

For $r > |c| + k_1$

$$\left| \oint_{\partial_1(r)} \frac{c+K(z)}{zf(z)} dz \right| \leq \frac{|c|+k_1}{r-|c|-k_1}\pi,$$

$$\left| \oint_{\partial_1(r)} \frac{K'(z)}{f(z)} dz \right| \leq \frac{r}{r-|c|-k_1} \int_{\theta=-\pi/2}^{\theta=\pi/2} \left|K'\left(re^{i\theta}\right)\right| d\theta$$

since $|K(z)| \leq k_1$ for $\operatorname{Re} z \geq 0$. Thus,

$$\lim_{r\to+\infty} \oint_{\partial_1(r)} \frac{c+K(z)}{zf(z)} dz = 0.$$

Now $K'(z) = \int_0^{+\infty} ae^{-za} da$ and $\lim_{r\to+\infty} |K'(re^{i\theta})| = 0$ for $-\frac{\pi}{2} < \theta < \frac{\pi}{2}$. By the dominated convergence theorem $\lim_{r\to+\infty} \int_{\theta=-\pi/2}^{\theta=\pi/2} |K'(re^{i\theta})| d\theta = 0$ and hence

$$\lim_{r\to+\infty} \oint_{\partial_1(r)} \frac{K'(z)}{f(z)} dz = 0.$$

It follows that

$$\lim_{r\to+\infty} \oint_{\partial_1(r)} \frac{f'(z)}{f(z)} dz = \lim_{r\to+\infty} \oint_{\partial_1(r)} \frac{1}{z} dz = \pi i.$$

Thus, the number of roots ν of $f(z)$ in the right half plane $\operatorname{Re} z \geq 0$ is

$$\nu = \lim_{r\to+\infty} \nu(r) = \frac{1}{2} + \frac{1}{2\pi} \lim_{r\to+\infty} \left(\arg f(-ir) - \arg f(ir)\right).$$

Since $f(\overline{z}) = \overline{f(z)}$ it follows that $\arg f(-ir) = -\arg f(ir)$ and $\nu = \frac{1}{2} - \frac{1}{\pi} \arg f(+i\infty)$, where $\arg f(+i\infty) \doteq \lim_{r\to+\infty} \arg f(ir)$.

Since $\lim_{r\to+\infty} \operatorname{Im} f(ir) = \lim_{r\to+\infty} (ir + c + K(ir)) = +i\infty$, it follows that $\arg f(+i\infty) = \frac{\pi}{2} - 2m\pi$ for some integer $m \in I(-\infty, +\infty)$ and the number of roots in the right half plane equals $2m$.

Bibliography

[1] J. D. Aber, Why don't we believe the models?, *Bull. Ecol. Soc. Amer.* 78 (1997), 232–233

[2] D. M. Adams and A. R. Ek, Optimizing the management of uneven-aged forest stands, *Can. J. For. Res.* 4 (1974), 274–287

[3] F. R. Adler, Coexistence of two types on a single resource in discrete time, *J. Math. Biol.* 28 (1990), 695–713

[4] W. G. Aiello, H. I. Freedman, and J. Wu, Analysis of a model representing stage-structured population growth with state-dependent time delay, *SIAM J. Appl. Math.* 52 (1992), 855–869

[5] W. C. Allee, *Animal Aggregations*, University of Chicago Press, Chicago, 1931

[6] W. C. Allee, *Cooperation Among Animals*, Henry Schuman, New York, 1951

[7] L. J. S. Allen, A density-dependent Leslie matrix model, *Math. Biosci.* 95 (1988), 179–187

[8] L. J. S. Allen, M. P. Moulton, and F. L. Rose, Persistence in an age-structured population for a patch-type environment, *Nat. Res. Modeling* 4 (1990), 197–214

[9] K. T. Alligood, T. D. Sauer, and J. A. Yorke, *Chaos: An Introduction to Dynamical Systems*, Springer-Verlag, Berlin, 1997

[10] O. Arino, D. E. Axelrod, and M. Kimmel (eds.), *Mathematical Population Dynamics*, Lecture Notes in Pure and Appl. Math. 131, Dekker, New York, 1991

[11] M. Artzrouni, On the dynamics of a population subject to slowly changing vital rates, *Math. Biosci.* 80 (1986), 265–290

[12] P. Auger, Stability of interacting populations with class-age distributions, *J. Theor. Biol.* 112 (1985), 595–605

[13] H. T. Banks, Some remarks on estimation techniques for size-structured population models, *Frontiers in Mathematical Biology* (S. A. Levin, ed.), Lecture Notes in Biomath. 100, Springer-Verlag, Berlin, 1995, 609–623

[14] H. T. Banks, L. W. Botsford, F. Kappel, and C. Wang, Modeling and estimation in size structured population models, *Mathematical Ecology,* World Scientific Publishing, Teaneck, NJ, 1988, 521–541

[15] H. T. Banks, L. W. Botsford, F. Kappel, and C. Wang, Estimation of growth and survival in size-structured cohort data: An application to larval striped bass (*Morone Saxatilis*), *J. Math. Biol.* 30 (1991), 125–150

[16] H. T. Banks and B. G. Fitzpatrick, Statistical methods for model comparison in parameter estimation problems for distributed systems, *J. Math. Biol.* 28 (1990), 501–527

[17] H. T. Banks and B. G. Fitzpatrick, Estimation of growth rate distributions in size structured population models, *Quart. Appl. Math.* 49 (1991), 215–235

[18] H. T. Banks, B. Fitzpatrick, and Y. Zhang, Estimation of distributed individual rates from aggregate population data, *Differential Equations and Applications to Biology and to Industry* (M. Martelli, K. Cooke, E. Cumberbatch, B. Tang, and H. Thieme, eds.), World Scientific Publishing, Singapore, 1996, 13–22

[19] H. T. Banks, F. Kappel, and C. Wang, Weak solutions and differentiability for size structured population models, *Internat. Ser. Numer. Math.* 100 (1991), 35–50

[20] H. T. Banks and K. Kunisch, *Estimation Techniques for Distributed Parameter Systems*, Birkhaüser, Boston, 1989

[21] T. H. Barr, Approximation for age-structured population models using projection methods, *Comput. Math. Appl.* 21 (1991), 17–40

[22] H. Barclay, Models of host-parasitoid interactions to determine the optimal instar of parasitization for pest control, *Natural Resource Modeling* 1 (1986), 81–103

[23] J. R. Beddington and C. A. Free, Age structure effects in predator-prey interactions, *Theoret. Population Biol.* 9 (1976), 15–24

[24] S. Bhattacharyya and C. G. Chakrabarti, On the dynamics of age-structured population, *Bull. Calcutta Math. Soc.* 83 (1991), 297–304

[25] J. Beddington, Age distribution and the stability of simple discrete time population models, *J. Theor. Biol.* 47 (1974), 65–74

[26] J. Beddington and C. A. Free, Age structure effects in predator-prey interactions, *Theoret. Population Biol.* 9 (1976), 15–24

[27] T. S. Bellows Jr. and M. P. Hassell, The dynamics of age-structured host-parasitoid interactions, *J. Anim. Ecol.* 57 (1988), 259–268

[28] J. Bengtsson, Smaller zooplankton species are not superior in exploitative competition: A comment on Persson, *Amer. Nat.* 129 (1987), 928–931

[29] M. O. Bergh and W. M. Getz, Stability of discrete age-structured and aggregated delay-difference population models, *J. Math. Biol.* 26 (1988), 551–581

[30] K. Blayneh, *A Hierarchically Size-Structured Population Model*, Ph.D. dissertation, University of Arizona, Tucson, AZ, 1995

[31] S. P. Blythe, R. M. Nisbet, and W. S. C. Gurney, Formulating population models with differential aging, *Population Biology* (H. I. Freeman and C. Strobeck, eds.), Lecture Notes in Biomath. 52 (1983), Springer-Verlag, Berlin, 133–140

[32] S. P. Blythe, R. M. Nisbet, and W. S. C. Gurney, The dynamics of population models with distributed maturation periods, *Theoret. Population Biol.* 25 (1984), 289–311

[33] L. W. Botsford, Optimal fishery policy for size-specific, density-dependent population models, *J. Math. Biol.* 12 (1981), 265–293

[34] L. W. Botsford, Individual state structure in population models, *Individual-Based Models and Approaches in Ecology* (D. L. DeAngelis and L. J. Gross, eds.), Chapman and Hall, New York, 1992, 213–236

[35] E. N. Boulanger, Small perturbations in nonlinear age-structured population equations, *J. Math. Biol.* 32 (1994), 521–533

[36] F. Brauer, Nonlinear age-dependent population growth under harvesting, *Int. J. Comput. Math. Appl.* 9 (1983), 345–352

[37] F. Brauer and M. Zhien, Stability of stage-structured population models, *J. Math. Anal. Appl.* 126 (1987), 301–315

[38] M. Brokate, Pontryagin's principle for control problems in age-dependent population dynamics, *J. Math. Biol.* 23 (1985), 75–101

[39] J. L. Brooks and S. I. Dodson, Predation, body size, and composition of plankton, *Science* 150 (1965), 28–35

[40] J. Buongiorno and B. R. Michie, A matrix model of uneven-aged forest management, *Forest Sci.* 26 (1980), 609–625

[41] S. N. Busenberg and M. Iannelli, Separable models in age-dependent population dynamics, *J. Math. Biol.* 22 (1985), 145–173

[42] S. N. Busenberg and C. C. Travis, On the use of reducible-functional differential equations in biological models, *J. Math. Anal. Appl.* 89 (1982), 46–66

[43] G. J. Butler and G. S. K. Wolkowicz, A mathematical model of the chemostat with a general class of functions describing nutrient uptake, *SIAM J. Appl. Math.* 45 (1985), 138–151

[44] W. A. Calder III, *Size, Function, and Life History*, Harvard University Press, Cambridge, MA, 1984

[45] A. Calsina and J. Saldana, A model of physiologically structured population dynamics with a nonlinear individual growth rate, *J. Math. Biol.* 33 (1995), 335–364

[46] A. Calsina and J. Saldana, Asymptotic behaviour of a model of hierarchically structured population dynamics, *J. Math. Biol.* 35 (1997), 869–907

[47] Y. L. Cao, J. Fan, and T. C. Gard, The effects of state-dependent time delay on a stage-structured population growth model, *Nonlinear Anal.* 19 (1992), 95–105

[48] Y. L. Cao and H. I. Freedman, Oscillation and global attractivity in stage-structured population models, *Canad. Math. Bull.* 36 (1993), 129–138

[49] H. Caswell, A general formula for the sensitivity of population growth rate to changes in life history parameters, *Theoret. Population Biol.* 14 (1978), 215–230

[50] H. Caswell, Stable population structure and reproductive value for populations with complex life cycles, *Ecology* 63 (1982), 1223–1231

[51] H. Caswell, Optimal life histories and the maximization of reproductive value: A general theorem for complex life cycles, *Ecology* 63 (1982), 1218–1222

[52] H. Caswell, Life history theory and the equilibrium status of populations, *Amer. Nat.* 120 (1982), 317–339

[53] H. Caswell, Life cycle models for plants, *Lectures on Mathematics in the Life Sciences* 18 (1986), 171–233

[54] H. Caswell, Approaching size and age in matrix population models, *Size-Structured Populations* (B. Ebenman and L. Persson, eds.), Springer-Verlag, Berlin, 1988, 85–105

[55] H. Caswell, *Matrix Population Models*, Sinauer Associates, Inc. Publishers, Sunderland, MA, 1989

[56] H. Caswell and A. M. John, From the individual to the population in demographic models, *Individual-Based Models and Approaches in Ecology* (D. L. DeAngelis and L. J. Gross, eds.), Chapman and Hall, New York, 1992, 36–61

[57] W.-L. Chan and B.-Z. Guo, Global behaviour of age-dependent logistic population models, *J. Math. Biol.* 28 (1990), 225–235

[58] B. Charlesworth, Natural selection in age-structured populations, *Lectures on Mathematics in the Life Sciences* 8 (1976), 69–87

[59] B. Charlesworth, *Evolution in Age Structured Populations*, Cambridge University Press, Cambridge, MA, 1980

[60] M. Chipot, On the equations of age-dependent population dynamics, *Arch. Rational Mech. Anal.* 82 (1983), 13–26

[61] M. Chipot and L. Edelstein, A mathematical theory of size distributions in tissue culture, *J. Math. Biol.* 16 (1983), 115–130

[62] B. Cipra, Chaotic bugs make the leap from theory to experiment, *SIAM News*, Vol. 30, No. 6 (July/August, 1997)

[63] A. J. Coale, *The Growth and Structure of Human Populations: A Mathematical Investigation*, Princeton University Press, Princeton, NJ, 1972

[64] J. E. Cohen, Ergodicity of age structure in populations with Markovian vital rates I: Countable states, *J. Amer. Statist. Assoc.* 71 (1976), 335–339

[65] J. E. Cohen, Ergodicity of age structure in populations with Markovian vital rates II: General states, *Adv. in Appl. Probab.* 9 (1977), 18–37

[66] J. E. Cohen, Ergodicity of age structure in populations with Markovian vital rates III: Finite-state moments and growth rate; An illustration, *Adv. in Appl. Probab.* 9 (1977), 462–475

[67] J. E. Cohen, Ergodic theorems in demography, *Bull. Amer. Math. Soc.* 1 (1979), 275–295

[68] B. D. Coleman, On the growth of populations with narrow spread in reproductive age. I. General theory and examples, *J. Math. Biol.* 6 (1978), 1–19

[69] C. S. Coleman and J. C. Frauenthal, Satiable egg eating predators, *Math. Biosci.* 63 (1983), 99–119

[70] D. Cooke and J. A. Leon, Stability of population growth determined by 2×2 Leslie matrix with density-dependent elements, *Biometrics* 32 (1976), 435–442

[71] R. F. Costantino and R. A. Desharnais, *Population Dynamics and the 'Tribolium' Model: Genetics and Demography*, Monographs on Theoretical and Applied Genetics, Vol. 13, Springer-Verlag, Berlin, 1991

[72] R. F. Costantino, J. M. Cushing, B. Dennis, and R. A. Desharnais, Experimentally induced transitions in the dynamics behaviour of insect populations, *Nature* 375 (May 18, 1995), 227–230

[73] R. F. Costantino, R. A. Desharnais, J. M. Cushing, and B. Dennis, Chaotic dynamics in an insect population, *Science* 275 (January 17, 1997), 389–391

[74] R. F. Costantino, J. M. Cushing, B. Dennis, R. A. Desharnais, and S. M. Henson, Resonant population cycles in alternating habitats, *Bull. Math. Biol.* 60 (1998), 247–275

[75] S. M. Henson, R. F. Costantino, J. M. Cushing, B. Dennis, and R. A. Desharnais, *Multiple Attractors, Saddles and Population Dynamics in Periodic Habitats*, preprint

[76] K. M. Crowe, *A Discrete Size-Structured Competition Model*, Ph.D. dissertation, Interdisciplinary Program in Applied Mathematics, University of Arizona, Tucson, AZ, 1991

[77] K. M. Crowe, *Persistence in Asymptotically Autonomous Systems of Difference Equations*, Fields Institute Technical Report, Toronto, ON, Canada, 1993

[78] K. M. Crowe, A nonlinear ergodic theorem for discrete systems, *J. Math. Biol.* 32 (1994), 179–191

[79] K. M. Crowe, Nonlinear ergodic theorems and symmetric versus asymmetric competition, *Structured-Population Models in Marine, Terrestrial, and Freshwater Systems* (Tuljapurkar and Caswell, eds.), Chapman and Hall, New York, 1997, Chapter 17

[80] K. M. Crowe and J. M. Cushing, Optimal instar parasitization in a stage-structured host-parasitoid model, *Natural Resource Modeling* 8 (1994), 119–138

[81] P. Cull and A. Vogt, Mathematical analysis of the asymptotic behavior of the Leslie population matrix model, *Bull. Math. Biol.* 35 (1973), 645–661

[82] P. Cull and A. Vogt, The periodic limit for the Leslie model, *Math. Biosci.* 21 (1974), 39–54

[83] P. Cull and A. Vogt, The period of total population, *Bull. Math. Biol.* 38 (1976), 317–319

[84] J. M. Cushing, An operator equation and bounded solutions of integro-differential systems, *SIAM J. Math. Anal.* 6 (1975), 433–445

[85] J. M. Cushing, Six Lectures on Integrodifferential Equations in Population Dynamics, *Proc. CIME Session on Mathematics in Biology* II, Ciclo 1979, Florence, Italy

[86] J. M. Cushing, *Integrodifferential Equations and Delay Models in Population Dynamics*, Lecture Notes in Biomathematics 20, Springer-Verlag, Heidelberg, 1979

[87] J. M. Cushing, Bifurcation of time periodic solutions of the McKendrick equations with applications to population dynamics, *Comp. and Math. Appl.* 9 (1983), 459–478 (Reprinted in *Advances in Hyperbolic Differential Equations*, Vol. 1, M. Witten, ed., Pergamon Press, Elmsford, NY, 1983.)

[88] J. M. Cushing, Equilibria in structured populations, *J. Math. Biol.* 23 (1985), 15–39

[89] J. M. Cushing, Global branches of equilibrium solutions of the McKendrick equations for age-structured population growth, *Comp. and Math. Appl.* 11 (1985), 175–188 (Reprinted in *Advances in Hyperbolic Differential Equations*, Vol. 2, M. Witten, ed., Pergamon Press, Elmsford, NY, 1985.)

[90] J. M. Cushing, Periodic McKendrick equations for age-structured population growth, *Comp. and Math. Appl.* 12A (1986), 513–526

[91] J. M. Cushing, Equilibria in systems of interacting age-structured populations, *J. Math. Biol.* 24 (1987), 627–649

[92] J. M. Cushing, Oscillatory population growth in periodic environments, *Theoret. Population Biol.* 30 (1987), 289–308

[93] J. M. Cushing, The Allee effect in age-structured population dynamics, *Mathematical Ecology* (S. A. Levin, T. Hallam, eds.), World Scientific Publishing, Teaneck, NJ, 1988, 479–505

[94] J. M. Cushing, A competition model for size-structured species, *SIAM J. Appl. Math.* 49 (1989), 838–858

[95] J. M. Cushing, A strong ergodic theorem for some nonlinear matrix models for the dynamics of structured populations, *Natural Resource Modeling* 3 (1989), 331–375

[96] J. M. Cushing, Some competition models for size-structured populations, *Roc. Mount. J. of Math.* 20 (1990), 879–897

[97] J. M. Cushing, Some delay models for juvenile vs. adult competition, *Differential Models in Biology, Epidemiology, and Ecology* (S. Busenberg and M. Matelli, eds.), Springer-Verlag, Berlin, 1991, 177–188

[98] J. M. Cushing, Competing size-structured species, *Mathematical Population Dynamics* (O. Arino, D.E. Axelrod, and M. Kimmel, eds.), Marcel Dekker, Inc., NY, 1991, Chapter 3

[99] J. M. Cushing, A simple model of cannibalism, *Math. Biosci.* 107 (1991), 47–72

[100] J. M. Cushing, A size-structured model for cannibalism, *Theoret. Population Biol.* 42 (1992), 347–361

[101] J. M. Cushing, A discrete model of competing size-structured species, *Theoret. Population Biol.* 41 (1992), 372–387

[102] J. M. Cushing, Nonlinear Matrix Models for Structured Populations, *Mathematical Population Dynamics: Analysis of Heterogeneity, Volume* III: *Mathematical Methods and Modelling of Data* (O. Arino, D. Axelrod, M. Kimmel, and M. Langlais, eds.), Wuerz Publishing Ltd, Winnipeg, MB, Canada, 1994, Chapter 26

[103] J. M. Cushing, Oscillations in age-structured population models with an Allee effect, *J. Comp. Appl. Math.* 52 (1994), 71–80

[104] J. M. Cushing, The dynamics of hierarchical age-structured populations, *J. Math. Biol.* 32 (1994), 705–729

[105] J. M. Cushing, Systems of difference equations and structured population dynamics, *Proc. of the First Int. Conf on Difference Equations* (S. N. Elaydi, J. R. Graef, G. Ladas, and A. C. Peterson, eds.), Gordon and Breach Publishers, Amsterdam, 1995, 123–132

[106] J. M. Cushing, Structured population dynamics, *Frontiers in Mathematical Biology* (S. A. Levin, ed.), Lecture Notes in Biomath. 100, Springer-Verlag, Berlin, 1995, 280–295

[107] J. M. Cushing, Periodically forced nonlinear systems of difference equations, *J. Difference Equations and Applications*, 3 (1998), 547–561

[108] J. M. Cushing and M. Saleem, A predator-prey model with age structure, *J. Math. Biol.* 14 (1982), 231–250

[109] J. M. Cushing and M. Saleem, A competition model with age structure, *Lecture Notes in Biomath.* 54 (1982), 178–192

[110] J. M. Cushing and J. Li, On Ebenman's model for the dynamics of a population with competing juveniles and adults, *Bull. Math. Biol.* 51 (1989), 687–713

[111] J. M. Cushing and J. Li, Juvenile versus adult competition, *J. Math. Biol.* 29 (1991), 457–473

[112] J. M. Cushing and J. Li, Intra-specific competition and density dependent juvenile growth, *Bull. Math. Biol.* 54 (1992), 503–529

[113] J. M. Cushing and J. Li, The dynamics of a size-structured intraspecific competition model with density dependent juvenile growth rates, *Individual Based Modeling: Concepts and Models* (D. DeAngelis and L. J. Gross, eds.), Routledge, Chapman, and Hall, New York, 1992, 112–125

[114] J. M. Cushing and J. Li, Oscillations caused by cannibalism in a size-structured population model, *Canad. Appl. Math. Quart.* 3 (1995), 155–172

[115] J. M. Cushing and Z. Yicang, The net reproductive value and stability in matrix population models, *Nat. Res. Mod.* 8 (1994), 297–333

[116] J. M. Cushing, B. Dennis, R. A. Desharnais, and F. R. Costantino, An interdisciplinary approach to understanding nonlinear ecological dynamics, *Ecol. Modeling* 91 (1996), 111–119

[117] J. M. Cushing, B. Dennis, R. A. Desharnais, and R. F. Costantino, Moving toward an unstable equilibrium: Saddle nodes in population systems, *J. Anim. Ecol.* 67 (1998), 298–306

[118] D. L. DeAngelis, D. K. Cox, and C. C. Coutant, Cannibalism and size dispersal in young-of-the-year large mouth bass: Experiment and model, *Ecol. Modelling* 8 (1979), 133–148

[119] D. L. DeAngelis and L. J. Gross (eds.), *Individual-Based Models and Approaches in Ecology*, Chapman and Hall, New York, 1992

[120] D. L. DeAngelis and J. S. Mattice, Implications of a partial differential equation cohort model, *Math. Biosci.* 47 (1979), 271–285

[121] D. L. DeAngelis, K. A. Rose, and M. A. Huston, Individual-oriented approaches to modeling ecological populations and communities, *Frontiers in Mathematical Biology* (S. A. Levin, ed.), Lecture Notes in Biomath. 100, Springer-Verlag, Berlin, 1995, 390–410

[122] D. L. DeAngelis, L. J. Svoboda, S. W. Christensen, and D. S. Vaughan, Stability and return times of Leslie matrices with density-dependent survival: Applications to fish populations, *Ecol. Modelling* 8 (1980), 149–163

[123] G. di Blasio, Nonlinear age-dependent population growth with history-dependent birth rate, *Math. Biosci.* 46 (1979), 279–291

[124] G. di Blasio, M. Iannelli, and E. Sinestrari, Approach to equilibrium in age structured populations with an increasing recruitment process, *J. Math. Biol.* 13 (1982), 371–382

[125] K. Deimling, Equilibria of an age-dependent population model, *Nonlinear Differential Equations: Invariance, Stability and Bifurcation* (P. de Mottoni and L. Salvadori, eds.), Academic Press, New York, 1981

[126] L. Demetrius, Primitivity conditions for growth matrices, *Math. Biosci.* 12 (1971), 53–58

[127] L. Demetrius, Natural selection and age-structured populations, *Genetics* 79 (1974), 535–544

[128] L. Demetrius, Demographic parameters and natural selection, *Proc. Nat. Acad. Sci.* 71 (1974), 4645–4647

[129] W. R. DeMott and W. C. Kerfoot, Competition among cladocerans: Nature of the interaction between *Bosmina* and *Daphnia*, *Ecology* 63 (1982), 1949–1966

[130] B. Dennis, Allee effects: Population growth critical density, and the chance of extinction, *Nat. Res. Modeling* 3 (1989), 481–538

[131] B. Dennis, P. O. Munholland, and J. M. Scott, Estimation of growth and extinction parameters for endangered species, *Ecol. Monogr.* 61 (1991), 115–143

[132] B. Dennis, R. A. Desharnais, J. M. Cushing, and R. F. Costantino, Nonlinear demographic dynamics: Mathematical models, statistical methods, and biological experiments, *Ecol. Monogr.* 65 (1995), 261–281

[133] B. Dennis, R. Desharnais, J. M. Cushing, and R. F. Costantino, Transitions in population dynamics: Equilibria to periodic cycles to aperiodic cycles, *J. Anim. Ecol.* 66 (1997), 704–729

[134] A. M. de Roos, Numerical methods for structured population models: The escalator boxcar train, *Numer. Methods Partial Differential Equations* 4 (1988), 173–195

[135] A. M. de Roos, J. A. J. Metz, E. Evers, and A. Peipoldt, A size dependent predator-prey interaction: Who pursues whom?, *J. Math. Biol.* 28 (1990), 609–643

[136] A. M. de Roos, O. Diekmann, and J. A. J. Metz, Studying the dynamics of structured population models: A versatile technique and its application to Daphnia, *Amer. Nat.* 139 (1992), 123–147

[137] R. A. Desharnais and J. E. Cohen, Life not lived due to disequilibrium in heterogeneous age-structured populations, *Theoret. Population Biol.* 29 (1986), 385–406

[138] R. A. Desharnais and L. Liu, Stable demographic limit cycles in laboratory populations of *Tribolium castaneum*, *J. Anim. Ecol.* 56 (1987), 885–906

[139] G. di Blasio, M. Iannelli, and E. Sinestrari, Approach to equilibrium in age structured populations with an increasing recruitment process, *J. Math. Biol.* 13 (1982), 371–382

[140] G. Di Cola and F. Nicoli, Parameter estimation in age-structured population dynamics, *Riv. Mat. Univ. Parma* (4) 9 (1983), 213–222

[141] O. Diekmann, The stable size distribution: An example in structured population dynamics, *Lecture Notes in Biomath.* 54 (1982), 90–96

[142] O. Diekmann, The dynamics of structured populations: Some examples, *Mathematics in Biology and Medicine* (V. Capasso, E. Grosso, and S. L. Paveri-Fontana, eds.), Lecture Notes in Biomath. 57, Springer-Verlag, Berlin, 1985, 7–18

[143] O. Diekmann, M. Gyllenberg, J. A. J. Metz, and H. Thieme, The "cumulative" formulation of (physiologically) structured population models, *Evolution Equations, Control Theory and Biomathematics* (Ph. Clément and G. Lumer, eds.), Lecture Notes in Pure and Appl. Math. 155, Marcel Dekker, New York, 1994, 145–154

[144] O. Diekmann, H. A. Lauwerier, T. Aldenberg, and J. A. J. Metz, Growth, fission and the stable size distribution, *J. Math. Biol.* 18 (1983), 135–148

[145] O. Diekmann and H. Metz, Exploring linear chain trickery for physiologically structured populations, *CWI Quarterly* 2 (1989), 3–14

[146] O. Diekmann and J. A. J. Metz, On the reciprocal relationship between life histories and population dynamics, *Frontiers in Mathematical Biology* (S. A. Levin, ed.), Lecture Notes in Biomath. 100, Springer-Verlag, Berlin, 1995, 263–279

[147] O. Diekmann, R. M. Nisbet, W. S. C. Gurney, and F. van den Bosch, Simple mathematical models for cannibalism: A critique and a new approach, *Math. Biosci.* 78 (1986), 21–46

[148] S. I. Dodson, Zooplankton competition and predation: An experimental test of the size-efficiency hypothesis, *Ecology* 55 (1974), 605–613

[149] J. Douglas Jr. and F. A. Milner, Numerical methods for a model of population dynamics, *Calcolo* 24 (1987), 247–254

[150] W. Ebeling, A. Engel, and V. G. Mazenko, Modeling of selection processes with age-dependent birth and death rates, *BioSystems* 19 (1986), 213–221

[151] B. Ebenman, Niche difference between age classes and intraspecific competition in age-structured populations, *J. Theor. Biol.* 124 (1987), 25–33

[152] B. Ebenman, Competition between age classes and population dynamics, *J. Theor. Biol.* 131 (1988), 389–400

[153] B. Ebenman, Dynamics of age- and size-structured populations: Intraspecific competition, *Size-Structured Populations: Ecology and Evolution* (B. Ebenman, L. Persson, eds.), Springer-Verlag, Berlin, 1988, 127–139

[154] B. Ebenman and L. Persson, *Size-Structured Populations: Ecology and Evolution*, Springer-Verlag, Berlin, 1989

[155] L. Edelstein-Keshet, Mathematical theory for plant-herbivore systems, *J. Math. Biol.* 24 (1986), 25–58

[156] S. N. Elaydi, *An Introduction to Difference Equations*, Springer-Verlag, New York, 1996

[157] R. H. Elderkin, Structured populations with nonlinear, nonlocally dependent dynamics: Basic theory, *J. Math. Anal. Appl.* 192 (1995), 392–412

[158] M. A. Elgar and B. J. Crespi, *Cannibalism: Ecology and Evolution Among Diverse Taxa*, Oxford University Press, Oxford, 1992

[159] Y. S. Fan and L. S. Chen, Periodic solutions and period-similarity of a growth model with age-structured population, *Systems Sci. Math. Sci.* 4 (1991), 148–157

[160] M. E. Fisher and B. S. Goh, Stability results for delayed-recruitment models in population dynamics, *J. Math. Biol.* 19 (1984), 147–156

[161] B. G. Fitzpatrick, Statistical tests of fit in estimation problems for structured population modeling, *Quart. Appl. Math.* 53 (1995), 105–128

[162] L. R. Fox, Cannibalism in natural populations, *Ann. Rev. Ecol. Syst.* 6 (1975), 87–106

[163] J. E. Franke and A.-A. Yakubu, Global attractors in competitive systems, *Nonlinear Anal.* 16 (1991), 111–129

[164] J. E. Franke and A.-A. Yakubu, Mutual exclusion versus coexistence for discrete competitive systems, *J. Math. Biol.* 30 (1991), 161–168

[165] J. E. Franke and A.-A. Yakubu, Pioneer exclusion in a one-hump discrete pioneer-climax competitive system, *J. Math. Biol.* 32 (1994), 771–787

[166] J. E. Franke and A.-A. Yakubu, Extinction in systems of bobwhite quail populations, *Canad. Appl. Math. Quart.* 3 (1995), 173–201

[167] J. E. Franke and A.-A. Yakubu, Dominance conditions in age-structured discrete competitive systems, *Proc. First Int. Conf. on Diff. Equations and Appl.* (S. N. Elaydi, J. R. Graef, G. Ladas, and A. C. Peterson, eds.), Gordon and Breach Publishers, New York, 1995, 197–211

[168] J. E. Franke and A.-A. Yakubu, Dominance conditions in age-structured discrete noncooperative systems, *J. Math. Biol.* 34 (1996), 442–454

[169] J. E. Franke and A.-A. Yakubu, Extinction and persistence of species in discrete competitive systems with a safe refuge, *J. Math. Anal. Appl.* 203 (1996), 746–761

[170] J. E. Franke and A.-A. Yakubu, Extinction of species in age-structured discrete noncooperative system, *J. Math. Biol.* 34 (1996), 442–454

[171] J. E. Franke and A.-A. Yakubu, Global asymptotic behavior and dispersion in age-structured, discrete competitive systems, to appear in *Canad. Appl. Math. Quart.* 5

[172] J. E. Franke and A.-A. Yakubu, Principles of competitive exclusion for discrete populations with reproducing juveniles and adults, *Proc. World Congress of Nonlinear Analysts*, 30 (1997), 1197–1206

[173] J. E. Franke and A.-A. Yakubu, Exclusionary population dynamics in size-structured, discrete competitive systems, *J. Differ. Equations Appl.*, in press

[174] J. C. Frauenthal, Some simple models of cannibalism, *Math. Biosci.* 63 (1983), 87–98

[175] H. I. Freedman and J. W.-H. So, Persistence in discrete semidynamical systems, *SIAM J. Math. Anal.* 20 (1989), 930–938

[176] G. Frobenius, Über matrizen aus nicht negativen elementen, *S.-B. Deutsch. Akad. Wiss. Berlin. Math-Nat. Kl.* (1912), 456–477

[177] T. Fujimoto and U. Krause, Asymptotic properties for inhomogeneous iterations of nonlinear operators, *SIAM J. Math. Anal.* 19 (1988), 841–853

[178] T. B. K. Gage, F. M. Williams, and J. B. Horton, Division synchrony and the dynamics of microbial populations: A size-specific model, *Theoret. Population Biol.* 26 (1986), 296–314

[179] F. R. Gantmacher, *The Theory of Matrices*, Vol. 2, Chelsea, New York, 1960

[180] J. Gerritsen, Size efficiency reconsidered: A general foraging model for free-swimming aquatic animals, *Amer. Nat.* 123 (1984), 450–467

[181] W. M. Getz, Optimal harvesting of structured populations, *Math. Biosci.* 44 (1979), 269–291

[182] W. M. Getz, Harvesting discrete nonlinear age and stage structured populations, *J. Optim. Theory Appl.* 57 (1988), 69–83

[183] W. M. Getz, A metaphysiological approach to modeling ecological populations and communities, *Frontiers in Mathematical Biology* (S. A. Levin, ed.), Lecture Notes in Biomath. 100, Springer-Verlag, Berlin, 1995, 411–442

[184] W. M. Getz and R. G. Haight, *Population Harvesting: Demographic Models of Fish, Forest, and Animal Resources*, Monographs in Population Biology 27, Princeton University Press, Princeton, NJ, 1989

[185] C. Godfray and M. Hassell, Chaotic beetles, *Science* (January 17, 1997), 323

[186] M. Golubitsky, E. B. Keeler, and M. Rothschild, Convergence of the age-structure: Applications of the projective metric, *Theoret. Population Biol.* 7 (1975), 84–93

[187] L. A. Goodman, On the sensitivity of the intrinsic growth rate to changes in the age-specific birth and death rates, *Theoret. Population Biol.* 2 (1971), 339–354

[188] K. Gopalsamy, Age specific coexistence in two species competition, *Math. Biosci.* 61 (1982), 101–122

[189] G. Greiner, A typical Perron-Frobenius theorem with applications to an age-dependent population equation, *Infinite-dimensional Systems* (F. Kappel and W. Schappacher, eds.), Lecture Notes in Math. 1076, Springer-Verlag, Berlin, 1984, 86–100

[190] G. Greiner, J. A. P. Heesterbeek, and J. A. J. Metz, A singular perturbation theorem for evolution equations and time-scale arguments for structured population models, *Canad. Appl. Math. Quart.* 2 (1994), 435–459

[191] D. H. Griffel, Age dependent population growth, *IMA J. Appl. Math.* 17 (1976), 141–152

[192] J. Guckenheimer, G. Oster, and A. Ipaktchi, The dynamics of density dependent population models, *J. Math. Biol.* 4 (1977), 101–147

[193] J. Guckenheimer and P. Homes, *Nonlinear Oscillations, Dynamical Systems, and Bifurcations of Vector Fields*, Springer-Verlag, Berlin, 1983

[194] W. S. C. Gurney and R. M. Nisbet, Age and density-dependent population dynamics in static and variable environments, *Theoret. Population Biol.* 17 (1980), 321–344

[195] W. S. C. Gurney and R. M. Nisbet, The systematic formulation of delay-differential models of age or size structured populations, *Population Biology* (Freedman and Strobeck, eds.), Lecture Notes in Biomath. 52, Springer-Verlag, Berlin, 1983

[196] W. S. C. Gurney and R. M. Nisbet, Fluctuation periodicity, generation separation and the expression of larval competition, *Theoret. Population Biol.* 28 (1985), 150–180

[197] W. S. C. Gurney, R. M. Nisbet, and J. H. Lawton, The systematic formulation of tractable single species population models incorporating age structure, *J. Anim. Ecol.* 52 (1983), 479–495

[198] M. E. Gurtin and R. C. MacCamy, Nonlinear age-dependent population dynamics, *Arch. Rational Mech. Anal.* 54 (1974), 281–300

[199] M. E. Gurtin and R. C. MacCamy, Some simple models for nonlinear age-dependent population dynamics, *Math. Biosci.* 43 (1979), 199–211

[200] M. E. Gurtin and R. C. MacCamy, Population dynamics with age dependence, *Nonlinear Analysis and Mechanics*, Herrot-Watt Symposium III, Pitman, Boston, 1979

[201] M. E. Gurtin and D. S. Levine, On predator-prey interactions with predation dependent on age of prey, *Math. Biosci.* 47 (1979), 207–219

[202] M. E. Gurtin and L. F. Murphy, On the optimal harvesting of persistent age-structured populations, *J. Math. Biol.* 13 (1981), 131–148

[203] M. E. Gurtin and L. F. Murphy, On the optimal harvesting of age-structured populations: Some simple models, *Math. Biosci.* 55 (1981), 115–136

[204] M. Gyllenberg, Nonlinear age-dependent population dynamics in continuously propagated bacterial cultures, *Math. Biosci.* 62 (1982), 45–74

[205] M. Gyllenberg, Stability of a nonlinear age-dependent population model containing a control variable, *SIAM J. Appl. Math.* 43 (1983), 1418–1438

[206] M. Gyllenberg, G. Söderbacka, and S. Ericsson, Does migration stabilize local population dynamics? Analysis of a discrete metapopulation model, *Math. Biosci.* 118 (1993), 25–49

[207] M. Gyllenberg and G. F. Webb, A nonlinear structured population model of tumor growth with quiescence, *J. Math. Biol.* 28 (1990), 671–694

[208] K. P. Hadeler, Pair formation in age-structured populations, *Acta Appl. Math.* 14 (1989), 91–102

[209] J. W. Haefner, *Modeling Biological Systems: Principles and Applications*, Chapman and Hall, New York, 1996

[210] A. Haimovici, On the growth of a population dependent on ages and involving resources and pollution, *Math. Biosci.* 43 (1979), 213–237

[211] J. K. Hale, *Asymptotic Behavior of Dissipative Systems*, Mathematical Surveys and Monographs 25, AMS, Providence, RI, 1985

[212] D. J. Hall, S. T. Threlkeld, C. W. Burns, and P. H. Crowley, The size-efficiency hypothesis and the size structure of zooplankton communities, *Ann. Rev. Ecol. Syst.* 7 (1976), 177–208

[213] T. G. Hallam and S. M. Henson, Extinction in a structured predator-prey model with size-dependent predation, *Differential Equations and Applications to Biology and to Industry* (M. Martelli, K. Cooke, E. Cumberbatch, B. Tang, and H. Thieme, eds.), World Scientific, Singapore, 1996, 157–172

[214] T. G. Hallam, R. R. Lassiter, J. Li, and W. McKinney, Physiologically structured population models in risk assessment, *Biomathematics and Related Computational Problems*, Kluwer Academic Publishers, Dordrecht, 1988, 197–211

[215] K. B. Hannsgen and J. J. Tyson, Stability of the steady-state size distribution in a model of cell growth and division, *J. Math. Biol.* 22 (1985), 293–301

[216] A. Hastings, Simple models for age dependent predation, *Lecture Notes in Biomath.* 54, Springer-Verlag, Berlin, 1984, 114–119

[217] A. Hastings, Age-dependent predation is not a simple process I. Continuous time models, *Theoret. Population Biol.* 23 (1983), 347–362

[218] A. Hastings, Age-dependent predation is not a simple process II. Wolves, ungulates, and a discrete time model for predation on juveniles with a stabilizing tail, *Theoret. Population Biol.* 26 (1984), 271–282

[219] A. Hastings, Simple models for age-dependent predation, *Mathematical Ecology* (S. A. Levin and T. G. Hallam, eds.), Lecture Notes in Biomath. 54, Springer-Verlag, Berlin, 1984, 114–119

[220] A. Hastings, Cycles in cannibalistic egg-larval interactions, *J. Math. Biol.* 24 (1987), 651–666

[221] A. Hastings, Age-dependent dispersal is not a simple process: Density dependence, stability and chaos, *Theoret. Population Biol.* 41 (1992), 388–400

[222] A. Hastings and R. F. Costantino, Cannibalistic egg-larva interactions in Tribolium: An explanation for the oscillations in population numbers, *Amer. Nat.* 130 (1987), 36–52

[223] A. Hastings and R. F. Costantino, Oscillations in population numbers: Age-dependent cannibalism, *J. Anim. Ecol.* 60 (1991), 471–482

[224] A. Hastings and D. Wollkind, Age structure in predator-prey systems. A general model and a specific example, *Theoret. Population Biol.* 21 (1982), 44–56

[225] H. J. A. M. Heijmans, On the stable size distribution of populations reproducing by fission into two unequal parts, *Math. Biosci.* 72 (1984), 19–50

[226] H. J. A. M. Heijmans, Structured populations, linear semigroups and positivity, *Math. Z.* 191 (1986), 599–617

[227] S. M. Henson, Existence and stability of nontrivial periodic solutions of periodically forced discrete dynamical systems, *J. Differ. Equations Appl.* 2 (1996), 315–331

[228] S. M. Henson, Cannibalism can be beneficial even when its mean yield is less than one, *Theoret. Population Biol.* 51 (1977), 109–117

[229] S. M. Henson, *A Continuous, Age-Structured Insect Population Model*, preprint

[230] S. M. Henson and J. M. Cushing, Hierarchical models of intra-specific competition: Scramble versus contest, *J. Math. Biol.* 34 (1996), 755–772

[231] S. M. Henson and J. M. Cushing, Optimal parasitization in a size-structured host-parasitoid model, *Nonlinear World* 3 (1996), 1–18

[232] S. M. Henson and J. M. Cushing, The effect of periodic habitat fluctuations on a nonlinear insect population model, *J. Math. Biol.* 36 (1997), 201–226

[233] S. M. Henson and T. G. Hallam, Survival of the fittest: Asymptotic competitive exclusion in structured population and community models, *Nonlinear World* 1 (1994), 385–402

[234] J. Hofbauer, V. Hutson, and W. Jansen, Coexistence for systems governed by difference equations of Lotka-Volterra type, *J. Math. Biol.* 25 (1987), 553–570

[235] J. Hofbauer and J. W.-H. So, Uniform persistence and repellors for maps, *Proc. AMS* 107 (1989), 1137–1142

[236] F. Hoppensteadt, *Mathematical Theories of Populations: Demographics, Genetics and Epidemics*, Regional Conference Series in Applied Mathematics 20, SIAM, Philadelphia, PA, 1975

[237] F. Hoppensteadt, *Mathematical Methods of Population Biology*, Cambridge University Press, Cambridge, 1982

[238] J. W. Horweed and J. G. Shepherd, The sensitivity of age-structured populations to environmental variability, *Math. Biosci.* 57 (1981), 59–82

[239] F. Houllier and J.-D. Lebreton, A renewal-equation approach to the dynamics of stage-grouped populations, *Math. Biosci.* 79 (1986), 185–197

[240] Y. H. Hsieh, On the evolution of altruism in an age-structured population, *Math. Comput. Modelling* 11 (1988), 472–475

[241] X. C. Huang, An age-dependent population model and its operator, *Phys. D.* 41 (1990), 356–370

[242] T. P. Hughes and J. H. Connell, Population dynamics based on size or age? A reef-coral analysis, *Amer. Nat.* 129 (1987), 818–829

[243] M. Iannelli, Mathematical problems in the description of age structured populations, *Mathematics in Biology and Medicine* (V. Capasso, E. Grosso and S. L. Paveri-Fontana, eds.), Lecture Notes in Biomath. 57, Springer-Verlag, Berlin, 19–32

[244] M. Iannelli, *Mathematical Theory of Age-Structured Population Dynamics*, Applied Mathematics Monographs 7, Giardini Editori e Stampatori, Pisa, 1994

[245] J. Impagliazzo, Deterministic Aspects of Mathematical Demography, *Biomathematics* 13, Springer-Verlag, Berlin, 1980

[246] H. Inaba, A semigroup approach to the strong ergodic theorem of the multi-state stable population process, *Math. Pop. Studies* 1 (1988), 49–77

[247] H. Inaba, Weak ergodicity of population evolution processes, *Math. Biosci.* 96 (1989), 195–219

[248] U. Inprasit, *Equilibria in Size-Structured Population Models*, Ph.D. dissertation, University of Arizona, Tucson, AZ, 1996

[249] K. Ito, F. Kappel, and G. Peichl, A fully discretized approximation scheme for size-structured population models, *SIAM J. Numer. Anal.* 28 (1991), 923–954

[250] V. A. A. Jansen, R. M. Nisbet, and W. S. C. Gurney, Generation cycles in stage structured populations, *Bull. Math. Biol.* 52 (1990), 375–396

[251] A. L. Jensen, Leslie matrix models for fisheries studies, *Biometrics* 30 (1974), 547–551

[252] D. Jillson, Insect populations respond to fluctuating environments, *Nature* 288 (1980), 699–700

[253] J. N. Kapur, Age-structured population models with density dependence, *Bull. Can. Math. Soc.* 74 (1982), 207–215

[254] J. N. Kapur, Age-structured population models with density dependence, *Bull. Calcutta Math. Soc.* 74 (1982), 207–215

[255] P. Kareiva, Predicting and producing chaos, *Nature* 375 (May, 1995), 189–190

[256] W. O. Kermack and A. G. McKendrick, A contribution to the mathematical theory of epidemics, *Proc. Roy. Soc.* A 115 (1927), 700–721

[257] W. O. Kermack and A. G. McKendrick, Contributions to the mathematical theory of epidemics II. The problem of endemicity, *Proc. Roy. Soc.* A 138 (1932), 55–83

[258] W. O. Kermack and A. G. McKendrick, Contributions to the mathematical theory of epidemics III. Further studies on the problem of endemicity, *Proc. Roy. Soc.* A 141 (1933), 92–122

[259] N. Keyfitz, *Introduction to the Mathematics of Populations*, Addison–Wesley, Reading, MA, 1968

[260] N. Keyfitz and J. A. Beekman, *Demography through Problems*, Springer-Verlag, New York, 1984

[261] N. Keyfitz, *Applied Mathematical Demography*, Springer-Verlag, Berlin, 1985

[262] J. Knight, Boom time for beetles, *New Scientist* 156, No. 2110 (November, 1997), 19

[263] S. A. L. M. Kooijman, *Dynamic Energy Budgets in Biological Systems*, Cambridge University Press, Cambridge, 1993

[264] T. Kostova, Oscillations in age dependent population dynamics, *Differential Equations and Applications to Biology and to Industry* (M. Martelli, K. Cooke, E. Cumberbatch, B. Tang, and H. Thieme, eds.), World Scientific, Singapore, 1996, 253–260

[265] T. V. Kostova and N. H. Chipev, A model of the dynamics of intramolluscan trematode populations: Some problems concerning oscillatory behavior, *Comp. Math. and Appl.* 21 (1991), 1–15

[266] T. Kostova and F. Milner, Nonlinear age-dependent population dynamics with constant size, *SIAM J. Math. Anal.* 22 (1991), 129–137

[267] M. Kot, Discrete-time travelling waves: Ecological examples, *J. Math. Biol.* 30 (1992), 413–436

[268] M. Kot and W. M. Schaffer, Discrete-time growth-dispersal models, *Math. Biosci.* 80 (1986), 109–136

[269] S. V. Krishna and S. Bhattacharyya, Study of structured populations with quiescent state, *Differential Equations Dynam. Systems* 1 (1993), 325–340

[270] Y. Kuan, *Delay Differential Equations: With Applications in Population Dynamics*, Academic Press, Boston, 1993

[271] Y. Kuang and J. M. Cushing, Global stability in a nonlinear difference-delay equation model of flour beetle population growth, *J. Differ. Equations Appl.* 2 (1995), 31–37

[272] R. H. Lamberson, K. McKelvey, B. R. Noon, and C. Voss, A dynamic analysis of Northern Spotted Owl viability in a fragmented forest landscape, *Conservation Biology* 6 (1992), 505–512

[273] L. Lamberti and P. Vernole, Existence and asymptotic behaviour of solutions of an age structured population model, *Boll. Un. Mat. Ital. C* (5) 18 (1981), 119–139

[274] H. D. Landahl and B. D. Hansen, A three stage population model with cannibalism, *Bull. Math. Biol.* 37 (1975), 11–17

[275] R. Lande, Demographic models of the northern spotted owl (*Strix occidentalis caurina*), *Oecologia* 75 (1988), 601–607

[276] R. Lande and S. H. Orzack, Extinction dynamics of age-structured populations in a fluctuating environment, *Proc. Nat. Acad. Sci.* 85 (1988), 7418–7421

[277] H. L. Lanhaar, General population theory in the age-time continuum, *J. Franklin Inst.* 293 (1972), 199–214

[278] J. P. LaSalle, *The Stability of Dynamical Systems*, Regional Conference Series in Applied Mathematics 25, SIAM, Philadelphia, PA, 1976

[279] R. Law, A model for the dynamics of a plant population containing individuals classified by age and size, *Ecology* 64 (1983), 224–230

[280] L. P. Lefkovitch, The study of population growth in organisms grouped by stages, *Biometrics* 25 (1965), 309–315

[281] P. H. Leslie, On the use of matrices in certain population mathematics, *Biometrika* 33 (1945), 183–212

[282] P. H. Leslie, Some further notes on the use of matrices in population mathematics, *Biometrika* 35 (1948), 213–245

[283] S. A. Levin and C. P. Goodyear, Analysis of an age-structured fishery model, *J. Math. Biol.* 9 (1980), 245–274

[284] S. A. Levin, Age-structure and stability in multiple-age spawning populations, *Renewable Resource Management* (T. L. Vincent and J. M. Skowronski, eds.), Springer-Verlag, Berlin, 1981, 21–45

[285] D. S. Levine, On the stability of a predator-prey system with egg-eating predators, *Math. Biosci.* 56 (1981), 27–46

[286] D. S. Levine, Bifurcating periodic solutions for a class of age-structured predator-prey systems, *Bull. Math. Biol.* 45 (1983), 901–915

[287] D. S. Levine, Models of age-dependent predation and cannibalism via the McKendrick equation, *Comp. Math. Appl.* 9 (1983), 403–414

[288] E. G. Lewis, On the generation and growth of a population, *Sankhya* 6 (1942), 93–96

[289] S. J. Lewis, A note on the strong ergodic theorem of some discrete models, *J. Differ. Equations Appl.* 3 (1997), 55–65

[290] J. Li, Persistence and extinction in continuous age-structured population models, *Comp. Math. Appl.* 15 (1988), 511–523

[291] J. Li, Persistence in discrete age-structured population models, *Bull. Math. Biol.* 50 (1988), 351–366

[292] J. Li and T. G. Hallam, Survival in continuous structured population models, *J. Math. Biol.* 26 (1988), 421–433

[293] X. R. Li and Z. E. Ma, An improved model of age-structured population dynamics, *Math. Comput. Modelling* 20 (1994), 143–150

[294] L. Liu and J. E. Cohen, Equilibrium and local stability in a logistic matrix model for age-structured populations, *J. Math. Biol.* 25 (1987), 73–88

[295] M. Lloyd, Self regulation of adult numbers by cannibalism in two laboratory strains of flour beetles (*Tribolium castaneum*), *Ecology* 49 (1967), 245–259

[296] D. O. Logofet, *Matrices and Graphs, Stability Problems in Mathematical Ecology*, CRC Press, London, 1992, 552–564

[297] A. Lomnicki, *Population Ecology of Individuals*, Monographs in Population Biology 25, Princeton University Press, Princeton, NJ, 1988

[298] A. Lomnicki and S. Sedziwy, Do individual differences in resource intakes without monopolization casue population stability and persistence?, *J. Theor. Biol.* 136 (1989), 317–326

[299] M. Loreau, Competition between age classes, and the stability of stage-structured populations: A re-examination of Ebenman's model, *J. Theor. Biol.* 144 (1990), 567–571

[300] M. Loreau and W. Ebenhöh, Competitive exclusion and coexistence of species with complex life cycles, *Theoret. Population Biol.* 46 (1994), 58–77

[301] A. J. Lotka, The stability of the normal age distribution, *Proc. Nat. Acad. Sci.* 8 (1922), 339–345

[302] M. Lynch, Complex interactions between natural coexpoliters–*Daphnia* and *Ceriodaphnia, Ecology* 59 (1978), 552–564

[303] N. MacDonald, *Time Lags in Biological Models*, Lecture Notes in Biomathematics 27, Springer-Verlag, Berlin, 1978

[304] P. Marcati, Asymptotic behavior in age-dependent population dynamics with hereditary renewal law, *SIAM J. Math. Anal.* 12 (1981), 904–916

[305] P. Marcati, On the global stability of the logistic age-dependent population growth, *J. Math. Biol.* 15 (1982), 215–226

[306] J. E. Marsden and M. McCracken, *The Hopf Bifurcation and Its Applications*, AMS 19, Springer-Verlag, Berlin, 1976

[307] C. F. Martin, L. J. S. Allen, and M. Stamp, An analysis of the transmission of chlamydia in a closed population, *J. Differ. Equations Appl.* 2 (1996), 1–20

[308] S. Matucci, Existence, uniqueness and asymptotic behaviour for a multi-stage evolution problem of an age-structured population. *Math. Models Methods Appl. Sci.* 5 (1995), 1013–1041

[309] R. M. May, *Stability and Complexity in Model Ecosystems*, Princeton University Press, Princeton, NJ, 1973

[310] R. M. May, Biological populations with nonoverlapping generations: Stable points, stable cycles, and chaos, *Science* 186 (1974), 645–647

[311] R. M. May, Simple mathematical models with very complicated dynamics, *Nature* 261 (1976), 459–467

[312] R. M. May, G. R. Conway, M. P. Hassell, and T. R. E. Southwood, Time delays, density-dependence and single-species oscillations, *J. Anim. Ecol.* 43 (1974), 747–770

[313] R. M. May and G. F. Oster, Bifurcations and dynamic complexity in simple ecological models, *Amer. Nat.* 110 (1976), 573–599

[314] J. Maynard Smith, *Models in Ecology*, Cambridge University Press, Cambridge, 1975

[315] J. Maynard Smith, *Mathematical Ideas in Biology*, Cambridge University Press, Cambridge, 1977

[316] N. G. Medhin and M. Sambandham, Analytical and numerical investigation of age structured population, *Dynam. Systems Appl.* 2 (1993), 435–450

[317] A. G. McKendrick and M. K. Pai, The rate of multiplication of micro-organisms: A mathematical study, *Proc. Roy. Soc. Edinburgh* 31 (1910), 649–655

[318] A. G. McKendrick, Applications of mathematics to medical problems, *Proc. Edin. Math. Soc.* 44 (1926), 98–130

[319] J. N. McNair, Stability effects of a juvenile period in age-structured populations, *J. Thoer. Biol.* 137 (1989), 397–422

[320] J. N. McNair and C. E. Goulden, The dynamics of age-structured populations with a gestation period: Density-independent growth and egg ratio methods for estimating the birth rate, *Theoret. Population Biol.* 39 (1991), 1–29

[321] J. A. J. Metz, A. M. de Roos, and F. van den Bosch, Population models incorporating physiological structures: A quick survey of the basic concepts and an application to size-structured population dynamics in waterfleas, *Size-Structured Populations: Ecology and Evolution* (B. Ebenman and L. Persson, eds.), Springer-Verlag, Berlin, 1989

[322] J. A. J. Metz and A. M. de Roos, The role of physiologically structured population models within a general individual-based modeling perspective, *Individual-Based Models and Approaches in Ecology* (D. L. DeAngelis and L. J. Gross, eds.), Chapman and Hall, New York, 1992, 88–111

[323] J. A. J. Metz and O. Diekmann, *The Dynamics of Physiologically Structured Populations*, Lecture Notes in Biomath. 68, Springer-Verlag, Berlin, 1986

[324] J. A. J. Metz and O. Diekmann, *Exact Finite-Dimensional Representations of Models for Physiologically Structured Populations* I. *The Abstract Foundations of Linear Chain Trickery*, Lecture Notes in Pure and Appl. Math. 133, Dekker, New York, 1991

[325] R. K. Miller, Asymptotic stability and perturbations for linear Volterra integrodifferential systems, *Delay and Functional Differential Equations and Their Applications* (K. Schmit, ed.), Academic Press, New York, 1972

[326] F. A. Milner, Age structured populations with history dependent mortality and natality, *Calcolo* 30 (1993), 29–39

[327] W. W. Murdoch, R. M. Nisbet, S. P. Blythe, W. S. C. Gurney, and J. D. Reeve, An invulnerable age class and stability in delay-differential parasitoid-host models, *Amer. Nat.* 129 (1987), 264–282

[328] L. F. Murphy, A nonlinear growth mechanism in size structured population dynamics, *J. Theor. Biol.* 104 (1983), 493–506

[329] L. F. Murphy, Density dependent cellular growth in an age-structured colony, *Comput. Math. Appl.* 9 (1983), 383–392

[330] L. F. Murphy and S. J. Smith, Optimal harvesting of an age-structured population, *J. Math. Biol.* 29 (1990), 77–90

[331] R. M. Nisbet and W. S. C. Gurney, *Modelling Fluctuating Populations*, Wiley and Sons, New York, 1982

[332] R. M. Nisbet and W. S. C. Gurney, The systematic formulation of population models for insects with dynamically varying instar duration, *Theoret. Population Biol.* 23 (1983), 114–135

[333] R. M. Nisbet, W. S. C. Gurney, S. P. Blythe, and J. A. J. Metz, Stage structure models of populations with distinct growth and development processes, *IMA J. Math. Appl. Biol. Med.* 2 (1985), 57–68

[334] R. M. Nisbet, W. S. C. Gurney, E. McCauley, and W. W. Murdoch, Structured population models: A tool for linking effects at individual and population levels, *Biol. J. Linnean Soc.* 37 (1989), 79–99

[335] R. M. Nisbet and L. C. Onyiah, Population dynamic consequences of competition within and between age classes, *J. Math. Biol.* 32 (1994), 329–344

[336] Z. Nitecki, *Differentiable Dynamics*, MIT Press, Cambridge, MA, 1971

[337] P. M. North, A computer modelling study of the population dynamics of the screech owl (*Otus asio*), *Ecol. Model.* 30 (1985), 105–114

[338] D. O'Regan and S. J. Smith, The maximum principle for continuous age-structured populations with unconstrained harvest rates, *Dynam. Systems Appl.* 3 (1994), 189–206

[339] G. Oster, Modern modeling of continuum phenomena, *Lectures in Population Dynamics* (R. C. diPrima, ed.), AMS, Providence, RI, 1977, 149–190

[340] G. Oster and Y. Takahashi, Models for age-specific interactions in a periodic environment, *Ecol. Monogr.* 44 (1974), 483–501

[341] S. W. Pacala and J. Weiner, Effects of competitive asymmetry on a local density model of plant interference, *J. Theor. Biol.* 149 (1991), 165–179

[342] C. J. Pennycuick, R. M. Compton, and L. Beckingham, A computer model for simulating the growth of a population, or of two interacting populations, *J. Theor. Biol.* 18 (1968), 316–329

[343] O. Perron, Über matrizen, *Math. Ann.* 64 (1907), 248–263

[344] M. Pilant and W. Rundell, Age structured population dynamics. Inverse methods in action, *Inverse Probl. Theor. Imaging*, Springer-Verlag, Berlin, 1990, 122–129

[345] R. E. Plant and L. T. Wilson, Models for age structured populations with distributed maturation rates, *J. Math. Biol.* 23 (1986), 247–262

[346] G. A. Polis, The evolution and dynamics of intraspecific predation, *Ann. Rev. Ecol. Syst.* 12 (1981), 225–251

[347] G. A. Polis, Age structure component of niche width and intra-specific resource partitioning: Can age groups function as ecological species?, *Amer. Nat.* 123 (1984), 541–564

[348] J. H. Pollard, *Mathematical Models for the Growth of Human Populations*, Cambridge University Press, London, 1975

[349] T. Prout and F. McChensney, Competition among immatures affects their adult fertility: Population dynamics, *Amer. Nat.* 126 (1985), 521–558

[350] J. Prüss, Equilibrium solutions of age-specific population dynamics of several species, *J. Math. Biol.* 11 (1981), 65–84

[351] J. Prüss, On the qualitative behaviour of populations with age-specific interactions, *Comput. Math.`Appl.* 9 (1983), 327–339

[352] J. Prüss, Stability analysis for equilibria in age-specific population dynamics, *Nonlinear Anal.* 7 (1983), 1291–1313

[353] J. Prüss and W. Schappacher, Semigroup methods for age-structured population dynamics, *Jahrbuch Überblicke Mathematik*, Vieweg, Braunschweig, 1994, 74–90

[354] P. H. Rabinowitz, Some global results for nonlinear eigenvalue problems, *J. Funct. Anal.* 7 (1971), 487–513

[355] S. Rai and D. Brewer, Analysis of a nonlinear FDE for an age-structured population, *J. Natur. Geom.* 11 (1997) 57–74

[356] J. Reed and N. C. Stenseth, Evolution of cannibalism in an age-structured population, *Bull. Math. Biol.* 46 (1984), 371–377

[357] W. J. Reed, Recruitment variability and age structure in harvested animal populations, *Math. Biosci.* 65 (1983), 239–268

[358] E. Renshaw, *Modelling Biological Populations in Space and Time*, Cambridge University Press, Cambridge, 1991

[359] D. J. Rodriquez, Models of growth with density dependence in more than one stage, *Theoret. Population Biol.* 34 (1988), 93–119

[360] P. Rohani and D. J. D. Earn, Chaos in a cup of flour, *Trends in Ecol. and Evol.* 12 (1997), 171

[361] C. Rorres, Stability of an age specific population with density dependent fertility, *Theoret. Population Biol.* 10 (1976), 26–46

[362] C. Rorres, A nonlinear model of population growth in which fertility is dependent on birth rate, *SIAM J. Appl. Math.* 37 (1979), 423–432

[363] C. Rorres, Local stability of a population with density-dependent fertility, *Theoret. Population Biol.* 16 (1979), 283–300

[364] S. Rosenblat, Population models in a periodically fluctuating envrinoment, *J. Math. Biol.* 9 (1980), 23–36

[365] M. Rotenberg, Equilibrium and stability in populations whose interactions are age-specific, *J. Theor. Biol.* 54 (1975), 207–224

[366] S. I. Rubinov, Age-structured equations in the theory of cell populations, *Studies in Mathematical Biology* II (S. A. Levin, ed.), MAA, Washington, 1978

[367] W. Rundell, Determining the death rate for an age-structured population from census data, *SIAM J. Appl. Math.* 53 (1993), 1731–1746

[368] K. Saints, *Discrete and Continuous Models of Age-Structured Population Dynamics,* Senior thesis, Harvey Mudd College, Claremont, CA, 1987

[369] M. Saleem, Predator-prey relationships: Egg-eating predators, *Math. Biosci.* 65 (1983), 187–197

[370] M. Saleem, Egg-eating age-structured predators in interaction with age-structured prey, *Math. Biosci.* 70 (1984), 91–104

[371] M. Saleem, Predator-prey relationships: Indiscriminate predation, *J. Math. Biol.* 21 (1984), 25–34

[372] M. Saleem, On interacting age-structured populations, *Math. Student* 55 (1989), 232–240

[373] W. M. Schaffer, Selection for optimal life histories: The effects of age structure, *Ecology* 55 (1974), 291–303

[374] F. M. Scudo and J. R. Ziegler, *The Golden Age of Theoretical Ecology*: 1923–1940, Lecture Notes in Biomath. 22, Springer-Verlag, Berlin, 1978

[375] F. R. Sharpe and A. J. Lotka, A problem in age-distribution, *Phil. Magazine* 21 (1911), 435–438

[376] H. R. Siegismund, V. Loeschcke, and J. Jacobs, Intraspecific competition and components of niche width in age structured populations, *Theoret. Population Biol.* 37 (1990), 291–319

[377] J. A. L. Silva and T. G. Hallam, Compensation and stability in nonlinear matrix models, *Math. Biosci.* 110 (1992), 67–101

[378] S. D. Simmes, *Age Dependent Population Dynamics with Nonlinear Interactions*, Ph.D. dissertation, Carnegie-Mellon University, Pittsburgh, PA, 1978

[379] E. Sinestrari, Nonlinear age-dependent population growth, *J. Math. Biol.* 9 (1980), 331–345

[380] J. W. Sinko and W. Streifer, A new model for age-size structure of a population, *Ecology* 48 (1967), 910–918

[381] J. W. Sinko and W. Streifer, Applying models incorporating age-size structure of a population to *Daphnia, Ecology* 50 (1969), 608–615

[382] D. Smith and N. Keyfitz, *Mathematical Demography*, Springer-Verlag, New York, 1977

[383] H. L. Smith, Equivalent dynamics for a structured population model and a related functional-differential equation, *Rocky Mountain J. Math.* 25 (1995), 491–499

[384] H. L. Smith, Reduction of structured population models to threshold-type delay equations and functional-differential equations: A case study, *Math. Biosci.* 113 (1993), 1–23

[385] H. L. Smith, A structured population model and a related functional-differential equation: Global attractors and uniform persistence, *J. Dynam. Differential Equations* 6 (1994), 71–99

[386] H. L. Smith, A discrete size-structured model of microbial growth and competition in the chemostat, *J. Math. Biol.* 34 (1996), 734–754

[387] H. L. Smith and P. Waltman, *The Theory of the Chemostat: Dynamics of Microbial Competition*, Cambridge Studies in Mathematical Biology, Cambridge University Press, Cambridge, 1995

[388] J. L. Smith and D. J. Wollkind, Age structure in predator-prey systems: Intraspecific carnivore interaction, passive diffusion and the paradox of enrichment, *J. Math. Biol.* 17 (1983), 275–288

[389] R. H. Smith and R. Mead, Age structure and stability in models of prey-predator systems, *Theoret. Population Biol.* 6 (1974), 308–322

[390] S. J. Smith, Maximum sustainable yield of a density dependent sex-age-structured population model, *J. Math. Biol.* 30 (1992), 801–814

[391] P. E. Smouse and K. M. Weiss, Discrete demographic models with density-dependent vital rates, *Oecologia* 21 (1975), 205–218

[392] W. Streifer, Realistic models in population ecology, *Advances in Ecological Research*, Vol. 8 (A. Macfadyen, ed.), Academic Press, New York, 1974

[393] D. Sulsky, R. R. Vance, and W. I. Newman, Time delays in age-structured populations, *J. Theor. Biol.* 141 (1989), 403–422

[394] D. Sulsky, Numerical solution of structured population models. I. Age structure, *J. Math. Biol.* 31 (1993), 817–839

[395] D. Sulsky, Numerical solution of structured population models. II. Mass structure, *J. Math. Biol.* 32 (1994), 491–514

[396] J. H. Swart, The stability of a nonlinear, age dependent population model, as applied to a biochemical reaction tank, *Math. Biosci.* 80 (1986), 47–56

[397] K. E. Swick, Stability and bifurcation in age-dependent population dynamics, *Theoret. Population Biol.* 20 (1981), 80–100

[398] K. E. Swick, A nonlinear model for human population dynamics, *SIAM J. Appl. Math.* 40 (1981), 266–278

[399] H. R. Thieme, Well-posedness of physiologically structured population models for *Daphnia magna.* How biological concepts can benefit by abstract mathematical analysis, *J. Math. Biol.* 26 (1988), 299–317

[400] H. R. Thieme, Analysis of age-structured population models with an additional structure, *Mathematical Population Dynamics* (O. Arino, D. E. Axelrod, and M. Kimmel, eds.), Lecture Notes in Pure and Appl. Math. 131, Dekker, New York, 1991, 115–126

[401] R. W. Thompson, D. di Blasio, and C. Mendes, Predator-prey interactions: Egg-eating predators, *Math. Biosci.* 60 (1982), 109–120

[402] W. O. Tschumy, Competition between juveniles and adults in age-structured populations, *Theoret. Population Biol.* 21 (1982), 255–268

[403] H. Tong, *Nonlinear Time Series: A Dynamical System Approach*, Oxford University Press, Oxford, 1990

[404] S. L. Tucker and S. O. Zimmerman, A nonlinear model of population dynamics containing an arbitrary number of continuous structure variables, *SIAM J. Appl. Math.* 48 (1988), 549–591

[405] S. D. Tuljapurkar, Population dynamics in variable environments IV: Weak ergodicity in the Lotka equation, *J. Math. Biol.* 14 (1982), 221–230

[406] S. Tuljapurkar, Entropy and convergence in dynamics and demography, *J. Math. Biol.* 31 (1993), 253–271

[407] S. Tuljapurkar, Stochastic demography and life histories, *Frontiers in Mathematical Biology* (Simon A. Levin, ed.), Lecture Notes in Biomath. 100, Springer-Verlag, Berlin, 1994, 254–262

[408] S. Tuljapurkar and H. Caswell, *Structured-Population Models in Marine, Terrestrial, and Freshwater Systems*, Chapman and Hall, New York, 1997

[409] G. Uribe, *On the Relationship Between Continuous and Discrete Models for Size-Structured Population Dynamics*, Ph.D. dissertation, Interdisciplinary Program in Applied Mathematics, University of Arizona, Tucson, AZ, 1993

[410] M. B. Usher, A matrix approach to the management of renewable resources, with special reference to selection forests, *J. Appl. Ecol.* 3 (1966), 355–367

[411] M. B. Usher, A matrix approach to the management of renewable resources, with special reference to selection forests–two extensions, *J. Appl. Ecol.* 6 (1969), 347–348

[412] M. B. Usher, A matrix model for forest management, *Biometrics* 25 (1969), 309–315

[413] M. B. Usher, Developments in the Leslie matrix model, *Mathematical Models in Ecology* (J. M. R. Jeffers, ed.), Blackwell Scientific Publishers, London, 1972, 29–60

[414] M. B. Usher, Extensions to modes, used in renewable resource management, which incorporate an arbitrary structure, *J. Environ. Man.* 4 (1976), 123–140

[415] M. B. Usher and M. H. Williamson, A deterministic matrix model for handling the birth, death and migration processes of spatially distributed populations, *Biometrics* 26 (1970), 1–12

[416] F. van den Bosch and O. Diekmann, Interactions between egg-eating predator and prey: The effect of the functional response and of age structure, *IMA J. Math. Appl. Med. Biol.* 3 (1986), 53–69

[417] F. van den Bosch, A. M. de Roos, and W. Babriel, Cannibalism as a life boat mechanism, *J. Math. Biol.* 26 (1988), 619–633

[418] J. van Sickle, Analysis of a distributed-parameter population model based on physiological age, *J. Theor. Biol.* 64 (1977), 571–586

[419] R. R. Vance, W. I. Newman, and D. Sulsky, The demographic meanings of the classical population growth models of ecology, *Theoret. Population Biol.* 33 (1988), 199–225

[420] J. Vandermeer, Choosing category size in a stage projection matrix, *Oecologia* 32 (1978), 79–84

[421] E. Venturino, Age-structured predator-prey models, *Math. Mod.* 5 (1984), 117–128

[422] M. O. Vlad, The occurrence of time-persistent distribution functions in structured population dynamics, *J. Theor. Biol.* 126 (1987), 239–242

[423] M. O. Vlad, Separable models for age-structured population genetics, *J. Math. Biol.* 26 (1988), 73–92

[424] H. Von Foerster, Some remarks on changing populations, *The Kinetics of Cellular Proliferation* (F. Stohlman, ed.), Grune and Stratton, New York, 1959, 382–407

[425] P. Waltman, *Competition Models in Population Biology*, Regional Conference Series in Applied Mathematics 45, SIAM, Philadelphia, PA, 1983

[426] F. S. J. Wang, Stability of an age-dependent population, *SIAM J. Math. Anal.* 11 (1980), 683–689

[427] G. F. Webb, Nonlinear age-dependent population dynamics in L^1, *J. Integral Equations* 5 (1983), 309–328

[428] G. F. Webb, The semigroup associated with nonlinear age dependent population dynamics, *Int. J. Comp. Math. and Appl.* 9 (1983), 487–497

[429] G. F. Webb, *Theory of Nonlinear Age-Dependent Population Dynamics*, Marcel Dekker, New York, 1985

[430] G. F. Webb, Dynamics of populations structured by internal variables, *Math. Z.* 189 (1985), 319–335

[431] G. F. Webb, Logistic models of structured population growth, *Hyperbolic Partial Differential Equations*, III. *Comput. Math. Appl.* Part A 12 (1986), 527–539

[432] G. F. Webb, An operator-theoretic formulation of asynchronous exponential growth, *Trans. Amer. Math. Soc.* 393 (1987), 751–763

[433] G. H. Weiss, Equations for the age structure of growing populations, *Bull. Math. Biophys.* 30 (1968), 427–435

[434] P. A. Werner and H. Caswell, Population growth rates and age versus stage-distribution models for teasel (*Dipsacus sylvestris* Huds.), *Ecology* 58 (1977), 1103–1111

[435] E. E. Werner and J. F. Gilliam, The ontogenetic niche and species interactions in size-structured populations, *Ann. Rev. Ecol. Syst.* 15 (1984), 393–425

[436] S. Wiggins, *Introduction to Applied Nonlinear Dynamical Systems and Chaos*, Texts in Applied Mathematics 2, Springer-Verlag, Berlin, 1990

[437] A. Wikan and E. Mjolhus, Periodicity of 4 in age-structured population models with density-dependence, *J. Theor. Biol.* 173 (1995), 109–119

[438] S. A. Wissinger, Niche overlap and the potential for competition and intraguild predation between size-structured populations, *Ecology* 73 (1992), 1431–1444

[439] D. Wollkind, A. Hastings, and J. Logan, Age structured in predator-prey systems II. Functional response and stability and the paradox of enrichment, *Theoret. Population Biol.* 21 (1982), 57–68

[440] I. O. Woodward, Modelling population growth in state-grouped organisms: A simple extension of the Leslie model, *Austr. J. of Ecol.* 7 (1982), 389–394

[441] M. E. J. Woolhouse and R. Harmsen, A transition matrix model of the population dynamics of a two-prey-two-predator acarid complex, *Ecol. Model.* 39 (1987), 307–323

[442] Z. Yicang and J. M. Cushing, Stability conditions for equilibria of nonlinear matrix population models, *J. Differ. Equations Appl.*, in press

[443] J. Yuan, Periodic solutions of a nonlinear autonomous age-dependent model of single species population dynamics, *Nonlinear Anal.* 12 (1988), 1079–1086.

[444] B. Xu, *A Discrete Nonlinear Model of Stage-Structured Populations*, Ph.D. dissertation, Interdisciplinary Program in Applied Mathematics, University of Arizona, Tucson, AZ, 1995

[445] D. W. Zachmann and J. A. Logan, Mathematical stability analysis for an age-structured population with finite life span, *J. Theor. Biol.* 111 (1984), 131–148

[446] D. W. Zachmann, Unique solvability of an age-structured population model with cannibalism, *Rocky Mountain J. Math.* 17 (1987), 749–766

[447] A. A. S. Zaghrout and S. H. Attalah, Analysis of a model of stage-structured population dynamics growth with time state-dependent time delay, *Appl. Math. Comput.* 77 (1996), 185–194

[448] A. D. Ziebur, Age-dependent models of population growth, *Theoret. Population Biol.* 26 (1984), 315–319

[449] M. F. Alameddine, *Size-Structured Competition Models with Periodic Coefficients*, Ph.D. dissertation, Interdisciplinary Program in Applied Mathematics, University of Arizona, Tucson, AZ, 1993

[450] Z. M. Alawneh, *A Numerical Method for Solving Certain Nonlinear Integral Equations Arising in Age-Structured Population Dynamics*, Ph.D. dissertation, University of Arizona, Tucson, AZ, 1990

Index